*In Search of*
## Divine
Reality

# In Search of
# Divine
# Reality

## Science as a Source of Inspiration

### Lothar Schäfer

The UNIVERSITY of
ARKANSAS PRESS
*Fayetteville* *1997*

01  00  99  98  97    5  4  3  2  1

*Designed by A. G. Carter*

♾  The paper used in this publication meets the minimum
requirements of the American National Standard for Permanence
of Paper for Printed Library Materials    Z39.48-1984.

LIBRARY OF CONGRESS CATALOGING-IN-PUBLICATION DATA

Schäfer, Lothar, 1939–
In search of divine reality : science as a source of inspiration / Lothar Schäfer.
p.    cm.
Includes bibliographical references and index.
ISBN 1-55728-458-X (cloth : alk. paper). — ISBN 1-55728-468-7 (pbk. : alk. paper)
1. Physics—Philosophy.  2. Transcendentalism.  3. Quantum theory.  4. God.
I. Title.
B67.S34    1997
110—dc21
97-2391
CIP

*To our children,*
*Nicole and Nathalie,*
*with Chris and Scott*

# ACKNOWLEDGMENTS

Writing this book has given me unabated happiness. After struggling with its conception for decades—helplessly for most of the time—all of a sudden it wrote itself. Gabriele, companion of a lifetime, was the benign force that made the process possible. Words cannot express my deep gratitude for her gentle love and selfless support.

No author has ever had a more understanding and helpful reader than I have had in Miller Williams, and I thank him for his passionate involvement. Countless thanks are also due to Debbie Self, for her expert guidance; to Collis Geren, dean of the Graduate School at Fayetteville and untiring servant to his faculty; to Elizabeth Payne, director of honors at Fayetteville, and Bernie Madison, dean of Arts and Sciences, for their support; to Susan Newton, for her competent and incessant help with details, particularly some of the figures; to Khamis Siam, for introducing me to the wonderful members of the Kansas Community College Math and Science Conferences at Pittsburg State University; to Diane Blair, Jim Blair, David Dubbell, John Ewbank, John Harrison, Mike Lieber, Erich Paulus, and Peter Pulay, for their friendship and stimulating discussions; and last but always first, to all my students who patiently listened to my thoughts and helped them grow.

# CONTENTS

*The indivisible elementary particle of modern physics possesses the quality of taking up space in no higher measure than other properties, say color and strength of material. In its essence, it is not a material particle in space and time . . . The experiences of modern physics show us that atoms do not exist as simple material objects . . .*

—WERNER HEISENBERG

*To put the conclusion crudely—the stuff of the world is mind-stuff . . . and the substratum of everything is of mental character . . . Consciousness is not sharply defined, but fades into subconsciousness; and beyond that we must postulate something indefinite but yet continuous with our mental nature. This I take to be the world-stuff.*

—SIR ARTHUR STANLEY EDDINGTON

# PREFACE

In this book I share my conclusions from a lifelong search for evidence—from quantum science—of the existence of a transcendent part of physical reality, combining disciplinary thought from science, philosophy, religion, and ethics.

When modern science adopted objectivity and experimental testing as operational principles, the ensuing advances in the physical sciences first clashed with, and then destroyed, the worldview of the ruling religious mythologies. A general disorientation resulted and a schism evolved in society, disconnecting the world of facts from the world of values. In a disastrous way the covenant which all religious mythologies claimed existed between humanity and nature was broken, and it seemed to many that life itself was rendered meaningless in the process.

Thus it is an important development that twentieth-century science has discovered—in the quantum phenomena—a part of physical reality which, in contrast to the mechanical world of classical science, has all the attributes of a transcendent reality. Now the foundation of the material world is found to be non-material; the constituents of real things are found not to be real in the same way as the things that they make; and non-local, faster-than-light influences are found to pervade a universe whose nature is mind-like.

Based on these findings the book guides the reader to the proposition that *the quantum phenomena make it possible to establish a new covenant*—between the human mind and the mind-like background of the universe—one that provides a home again to the homeless and meaning to seemingly pointless life. In this new understanding of the world, the universe must be assumed to have a moral as well as a physical order, and facts and values derive, again, from a single source.

In addition to describing the suggested connections, this book will explain, outside of the mathematical context in which they evolved, the scientific concepts relevant for its purpose. Particular attention will be given to the *quantum phenomena;* that is, the elementary phenomena that are now the domain of quantum mechanics. To be comprehensible to the educated generalist and layman, any information was included in the appendices that seemed essential, even that which experts in a given field might find rather elementary.

The explanatory material presented in this book is not part of its creative effort but builds on the inspiring sources quoted in the list of references—in addition to the historic classics, particularly those by *Eccles, Gribbin, Heisenberg, Herbert, Küng, Magee, Margenau, Monod, Polkinghorne, Popper, Russell, and Stapp*—masterpieces who have left a lasting impression on this author. Their ideas and statements have served as a framework around which my own ideas and this book were able to evolve.

Every author has probably made the shocking discovery that, as the words rushed from the mind easily, at times fragments from other texts inadvertently blended in, studied years ago and long since cleared from active memory. Every possible effort has been made to quote, to the extent that they were remembered, the sources of the material used in this book. However, after teaching a course on the subject for decades, combining personal notes with abstracts from the quoted references, all collected over a large span of my life, I am afraid that the origins of many specific formulations have been lost from memory and cannot be found again. All of them are herewith referenced collectively, and the citations listed at the end of this book are acknowledged as sources in toto and with modesty; they are strongly recommended as continuing reading material. If proper acknowledgment for the use of any material is missing, I would greatly appreciate being informed.

Those encountering the phenomena of the quantum world for the first time should be told now that a different person will leave this book than first entered it, because this is meant to be a healing book. The mechanical lawfulness of nature, originally postulated by classical science, now emulated by all the pseudosciences, and mindlessly reiterated by an educational system that tries to be popular rather than instructive, is often depressing to us. The discovery of the quantum world is a liberating experience. Things are not as inescapable as conventional wisdom implies; there is room for the unexpected, for adventure, for what Kundera called the unbearable lightness of being.

The first step to be taken toward an enlightened view of the nature of the universe and our position in it consists of an analysis of the ways we interact with the external world: *what exactly can we know about physical reality and how do we go about getting to know it?* The first step in a search for Divine Reality, then, is the search for transcendental elements in human knowledge. This is the first part of the book—the *Knowledge Module*. Expanding on this, the second part is the *Reality Module*—the search for transcendent physical reality; and the third part, the *Human-Nature Module*—the search for transcendental aspects of human nature. The final part, then, presents the possible conclusions regarding Divine Reality.

*Fayetteville, in the spring*
*of the fifty-eighth year.*

*In Search of*
**Divine**
**Reality**

# INTRODUCTION

## *I.*

The search for Divine Reality is the search for that part of physical reality which is the source of those universal principles which we need to lead an enlightened and virtuous life but cannot establish by a process of reasoning nor by an experience of the external world. This book is an inventory of what contemporary physical science has to offer as evidence for the existence of divine reality.

Within the human framework, the search for such a reality is the search for the transcendental—specifically, for the transcendental elements in human knowledge, the transcendental parts of physical reality, and the transcendental aspects of the nature of human beings. By transcendent I mean all those aspects, elements, or conditions of our existence which are beyond our direct control, beyond the visible surface of things, and beyond justification by reasoning or verification by the experience of our senses. Thus the search for divine reality is intimately related to the *reality-self-cognition* problem complex; that is, the complex of problems that we face in trying to understand the nature of physical reality, the nature of human beings, and the nature of human knowledge. In a short form the RSC problem complex is this: *The nature of human knowledge is not what it feels like. The nature of physical reality is not what it looks like. And as to human nature, even though it is our own, it is the least understood of them all.*

## *II.*

The problem of knowledge is the problem of **Hume**, that *the principles used in establishing factual knowledge cannot establish themselves; induction cannot induce induction; there are no verifiable general statements; and the basis of science is non-scientific.* Hume's problem is the fact that the processes that we employ in deriving knowledge require principles of inference which are *non-rational* and *non-empirical* in the sense that they cannot be derived by logical analyses or verified by scientific tests. These principles include the principles of *causality, induction,* and the assumption of *object permanence.* They can be adequately termed *epistemic;* that is, having to do with knowledge. No observations will ever *prove* the uninterrupted existence and identity of things. No law of nature will ever be *verified* by the inductive evidence that may suggest it. For Hume, *the contrary of all matters of fact is always possible. That the future does not resemble the past is not less logical an assumption, or in worse agreement with any experience, than the thesis that it does. And as to causality, there is no single experience of any causal event. It is not a principle of nature, but a habit of the human mind.*

Because of Hume's problem the elements of knowledge are bastards—sprung from the union of illegitimate partners—and the basis of knowledge is threefold. It involves, first, *experience of reality;* that is, it has to agree with observable facts. Second, *employment of reason;* that is, it has to be reasonable. Third, it requires the *employment of non-verifiable epistemic principles,* such as the certainty of matters of fact, the validity of induction, causality, the lawfulness of nature, and the faith in the continued identity and permanence of things. None of these principles is rooted in human experience or reason, in the sense that no experiment or reasoning can prove that they are true. Rather, they form an element in knowledge that is beyond human control, as though they were derived from a higher order or superior reason, from a part of nature that transcends the visible foreground of things. Some unknown process has placed these principles in the human mind, and it is an interesting question to ask where they come from.

As a possible solution to this problem I suggest that *the epistemic principles are rooted in nature but not in the space-time mechanistic order of nature. They are principles of the human mind because the mind is sensitive to influences from a part of reality that lies beyond the realm of its direct experience and reason.* This is the significance of the search for evidence —from physical science—of the existence of a transcendent part of physical reality.

## III.

Searching for transcendental physical reality, we find it in the world of *quantum phenomena*—the elementary phenomena which are now the domain of quantum mechanics. At its elementary foundation, the basis of material things reveals itself as *non-material;* the components of real things are *not real in the same way as those things that they form are real;* local order is affected by *non-local, faster-than-light* phenomena; deterministic processes alternate with *expressions of choices* in creating the visible order of things; *observation creates reality* and *entities with mind-like properties* are discovered. Each of these statements is a violation of common sense, which is entangled with the visible surface of things, but the phenomena encountered in this book will make them plausible.

The *wave-particle duality* is the defining symbol for the nature of the quantum world. It denotes the property of elementary physical entities—particles like electrons, protons, neutrons, or atoms and molecules—which is to exist in states which evolve like waves when they are not observed, and like particles when observed. The wave-like states are typically extended in space, but contract abruptly to localized events or point-like particles when an observation is made. Thus, when elementary particles are observed they always appear as tiny masses, as little lumps of matter. In a characteristic way, when a particle encounters another the two will collide like billiard balls, bounce around and push each other. An elementary particle is always observed as a localized event; for example, as a tiny flash on a phosphor screen or as a vanishing dot on a photographic plate. In contrast, when not observed those same things act like waves—call them *quantum waves*—that are delocalized, spread out through extended regions of space. When waves encounter each other, they do not collide and push, but superimpose, interpenetrate, interfere, adding constructively or destructively, creating interference patterns—in some places precipitating into deep valleys, and in others surging to crests of variable heights. This is the wave-particle duality.

As it turns out, the nature of quantum waves is that of *probability waves,* and the visible order of the universe is the result of their interference. Quantum wave functions are probability functions in the sense that they provide probabilities for possible outcomes of experiments—or possible events—to occur. For example, the wave function of an electron can provide the probability to find it at a given point in space. Due to the wave nature of this thing we cannot know for sure where it is, but the chances of finding it are given by the probability function—its quantum wave.

In the view of Heisenberg, the probability functions are

thought to possess objective reality. They are considered to be *objectively existing tendencies or possibilities* for actual events to occur. They are a modern version of Aristotle's concept of *potentia*. According to Aristotle matter without form was not quite real, but had the potential to become real by being formed. In the same way, Heisenberg thought, *the quantum waves are not quite real— "between the idea of a thing and a real thing"—but have the potential to become real when an observation is made.*

Accordingly, in Heisenberg's ontology—similar to views expressed by von Neumann and Born—physical reality is assumed to be formed by two processes. In the first, when no observations are made, elementary physical systems, such as atoms, molecules, or collections of elementary particles, evolve into a *superposition of possibilities or tendencies,* in a wave-like state that constantly splits into numerous branches, each representing an alternatively possible event, or outcome of an experiment. This state contains all possible events, evolving in accordance with *deterministic laws of motion,* and this process continues until an observation is made. In this case—the second process—the transition from the *"possible"* to the *"actual"* takes place, as Heisenberg phrased it. An observation *"changes the probability function discontinuously; it selects from all possible events the actual one that has taken place."* The second process occurs in the form of a *quantum jump.* Seemingly ruled by nothing but chance, one of the possible events is abruptly selected and made the actual one. In the terminology of **Dirac**, a *"choice"* is made. As Aristotle put it: *Forms bring matter into reality.* In the language of quantum mechanics: *Observations bring tendencies into reality.*

It is obvious that the wave-like state of elementary systems in Heisenberg's ontology is not an image of the physical reality of our conscious experience. The superposition of tendencies that have *the potential* to be real, but *are not,* is different than the reality that springs from it. Thus, *the constituents of real things are not as real as the things that they make.* Furthermore, since a state is created in the second Heisenberg process that did not exist before, it is reasonable to say that *observation creates reality,* where the term "observation" denotes an interaction of the quantum system with an object in a state of ordinary reality.

Similarly, since the quantum probability waves are empty, carrying no mass or energy, just numerical relations—while nevertheless, the rules of their interference inexorably dictate the visible order of things—it is reasonable to say that *the basis of the material world is non-material.*

In addition, since the choice and actualization of a possible event in the second Heisenberg process may involve a sudden change of possibilities through extended regions of the universe, it is reasonable to say that the nature of physical reality is non-local. *That is, an observation made in one part of the universe may have an instantaneous, faster-than-light effect on the possibilities of a second observer a long distance away.*

Finally, probabilities for occurrences of actual events are a form of information, closer by nature to elements of the mind than to the material world. In fact, processes are now known in which physical systems *react observably to mere changes in the information* available on them. Normally, reactions to information are the prerogative of a mind. Thus, *entities with mind-like properties are found at the foundation of ordinary things.*


## IV.

Like the wave-particle duality for physical reality, the *mind-body duality* is the defining symbol of the nature of human reality and indicator of its transcendent elements.

In the view of human reality set forth by **Eccles**, the central problem is *The Human Mystery: "How has the materialist mechanism of biological evolution been able to bring forth beings with self-consciousness and human values? How can we explain the dualism of man's nature: body and mind, and the interaction between mind and brain?"* In search of the transcendental elements of human nature, Eccles finds them in three processes: *in the process of evolution, the morphogenesis of a self from a single cell, and the mystery of visual perception.*

The line of events that led to the evolution of human beings is so unlikely, depended so much on incredible contingencies, and our existence is so unpredictable from the previous stages that, to men like Eccles, the *impression of guidance* is unavoidable. To begin with, the evolutionary development of self-conscious beings needed a universe with exactly the properties that ours happens to have. Make a slight change in any one of the universal constants, and our metabolism will break down. Furthermore, in our evolution, in Eccles' words, *"dead-ends were avoided in a mysterious way, . . . in that the species in our line of development were in some ways un-fit, less efficient, less competitive, . . . and therefore forced to continue on a narrow path that led to human beings."* For example, the fish in our line of ancestry were poor swimmers who could not compete with the ocean racers and finally had to crawl ashore, to become amphibians. For unknown reasons the latter added five digits to their limbs—one of the most essential traits of human anatomy. Next in line was a humble reptile that barely survived the onslaught of the much more powerful competitors, the dinosaurs. But it was the latter that perished; the "weaker" continued on the line to human beings, first to mammals and on to primates. At the level of mammals, dead-ends like the *"impressive"* carnivores, herbivores, or whalefish again were avoided. In fact, competition on the ground floor was so fierce that our primate ancestors had to get out of it and climb into trees. Thus, while the big mouths eventually lost their five digits, the latter had to hold on to them for dear life.

In the same way, in Eccles' ontology the transcendent is immanent in the morphogenesis of an individual from a single cell. All the cells in the body of an individual contain the same genetic material, the same information—*a thought*—in the form of DNA. All cells develop by repeated divisions from a single, initial cell. And yet an incredible differentiation takes place every time a human being is formed, leading to such diversified materials as muscles, receptors, tissues, membranes, organs, bones, and so on. In the words of **Sheldrake**: *"To believe that this procedure occurs without a design, without a masterplan, automatically, guided by nothing but the laws of physics and chemistry, is equivalent to the belief that, after delivering a ton of bricks, cement, wood, and glass to a construction site, the new building will erect itself spontaneously."*

In passionate opposition to such views, the ontology of **Monod**: In his book *Chance and Necessity*, he defends the thesis that *there are no transcendent elements in the nature of human beings, but the human body is a chemical reactor and its processes ruled by nothing but the laws of physics and chemistry.* Against the argument from evolution, the verdict by Monod is that *evolution is a gigantic lottery. Blind chance can lead to anything, even vision.* Furthermore, the evolution of humankind is a singular event and any statements on its likelihood are meaningless. He maintains, against the argument from morphogenesis, that there is no masterplan in cell differentiation, no mystery, there is nothing but ordinary chemical reactions, such as *selective gene activation* and *intermolecular non-bonded interactions*. Thus, nothing transcendent is immanent in the development of human beings or the processes that sustain them. Rather, to Monod, *the search for the transcendent is the longing for the animist covenant.*

*Vitalism* is the term for all theories which assume the existence of a special life force, a *teleonomic principle* in Monod's terminology, which operates only in living organisms, but nowhere else. *Animism* is the belief that all objects have a soul and are, in a way, alive; it is the assumption that a *universal teleonomic principle* is active throughout the entire cosmos. The important function of animism is that it conveys to human beings the soothing feeling of a kinship with nature, of having a purpose, of being a part of a Grand Design. In Monod's views it provides a protective *"covenant between nature and man, a profound alliance outside which there seems to stretch only terrifying solitude."*

Among our traditions of thinking, the tradition of modern science is based on *objectivity*. It is the belief that nature is objective because it is ruled by laws and not by purposes. The laws of nature are inviolable, quantitative, inexorable, and general. They will not bend for a purpose. Mechanical processes are aimless, purposeless; they follow the laws of motion, but have no goal. In contrast, when a process is the expression of a *final cause*, the intent to reach a goal, no matter how, is the cause of an action. *Objective science by definition must exclude purpose from its description of nature.*

Through much of our evolutionary history life in a tribe was the basis of survival. Once the non-human part of the biosphere was controlled to the extent that it did not represent serious dangers anymore, strife between tribes, mortal warfare, that great human achievement, provided the pressures of selection. At that point *cohesion of the horde* became a vital principle.

For social animals—ants, termites—civilization is genetically coded. The requisite behavior is automatic. For human beings, social behavior is not programmed so, but is supported by culture. To establish and maintain cohesion of the group, explanations are needed. In the course of cultural evolution, these were mostly provided in terms of myths and religions. The explanations told people why things were the way they were and, on that basis, why they had to act the way they were supposed to act. Thus, knowledge and values were derived from the same source.

In this history Monod saw the origin of our craving for the transcendent, for explanations and purpose in life. We have this craving, he believed, because we are the descendants of animists. Since the explanations of social order during most of the course of evolution were animist, projecting human nature into the rest of the world, he believed that our craving includes the need for a universal teleonomic principle: *"Every living being is also a fossil. Within it, all the way down to the microscopic structure of its proteins, it bears the traces if not the stigmata of its ancestry."*

In this world order of myths and religions, the entrance of science—the reliance on objectivity as the sole source of true knowledge—was a disaster. By first testing and then proving false the explanations given by myths and religions for the order of the universe, science destroyed their authority as a source of knowledge and thus, without intending to do so, shattered the foundation of social stability and the authority of the accepted systems of values.

In this way science pushed man into the *"icy solitude"* of a universe that did not share in human character, did not care about human concerns, had no plan in mind for his future. Life was seemingly without meaning, adrift in space-time, going nowhere, and man came to realize that (Monod) *"like a gypsy, he lives on the boundary of an alien world; a world that is deaf to his music, and as indifferent to his hopes as it is to his suffering or his crimes."* This is the story of the broken covenant between man and nature, the evolution of a life that does not make any sense, the reason why science has never been admitted to the hearts of the masses.

Thus arose the "lie at the root of the sickness of spirit" of modern societies that consists of the desperate clinging to values and myths whose foundations have long since been destroyed by that same principle of objectivity that is eagerly employed at the base of technology. On the one hand we have the successful performance of objectivity as a principle of knowledge, and science as the basis of our power and survival. On the other hand, we adhere to the old animist myths, but now without any basis, to preserve coherence in society and to justify what is right and what is wrong. Monod's conclusion was that a radical reappraisal of the traditional values and the rejection of many of them was unavoidable.

*Such a reappraisal can now be given, but with the result that a life with values and meaning is not in conflict with the principle of objectivity as the sole source of true knowledge. It is not the rejection of traditional values that is needed now, but the construction of a new foundation on which they can rest.*

## V.

The last statement can be made with confidence because:

- It is possible to identify transcendent elements in human knowledge.
- It is possible to identify transcendent parts of physical reality.
- It is possible to identify transcendent elements in human nature.
- If the universe is non-local, as we now must think, we must assume that human beings are necessarily a part of the web of effective influences that form the basis of its nature.
- If the nature of the universe is mind-like, as we now must think, we must assume that it includes human beings and communicates with the human mind.

*This is the restoration of the covenant with nature.*

Since the epistemic principles cannot be derived from an experience of the physical reality nor from a process of reasoning, we have asked the question, "Where do they come from?" As a possible answer it was suggested that they are rooted in nature, but not in the visible order of nature. In the same way, the ethical principles are not derived from an experience of the physical reality, nor from a process of reasoning—so where *do* they come from? The answer is the same as before: Like the epistemic principles, the moral principles are principles of the mind. They are rooted in nature—that is, in the transcendent part of nature. They are principles of the human mind, because mind is a natural extension of the mind-like background of the universe. Mind gives these principles authority. Being a part of a higher order, mind establishes the certainty of identity, permanence, factuality, and causality and the authority of honesty, morality, and purpose. The former tell us what is true and false; the latter, what is good and bad.

If, as **Eddington** says, "the substratum of everything is of mental character . . . [which in some parts] rises to the level of consciousness, . . . and the universe is of the nature of a thought," then *it must be assumed that the universe rules spiritual matters as well as the laws of physics, and it is in human minds that the spiritual order of the universe rises to the level of morality.* Perceived in this way, the epistemic principles are manifestations of the physical order of the universe; the moral principles, of its spiritual order. This is the common ground of facts and values and the restoration of the covenant with nature.

We cannot know what the mentality of the universe is like, but we know that, at the human level, it expresses itself in the feeling of consciousness. We cannot know what the spirituality of the universe is like, but we know that, at the human level, it expresses itself in the feeling of morality. Thus, the three fundamental questions of our existence—*what can I know, what should I do, what can I hope for*—find their natural answers: I can know what my senses, my reason, and the epistemic principles allow me to know. I should do what the ethical principles of the mind tell me is in harmony with the order of the universe. I can hope that the nature of the universe is mind-like and forms a covenant with my own mind as its natural extension.

Clearly, we need the covenant with nature, but not because we are the descendants of animists. Rather, because of the need of the mind to be in touch with what is akin to it in nature and to assure its subtle support.

Our music is the music of the universe, and Mozart's music is—as **Küng** wrote—"a touch of metaphysics." Evolution is the realization of universal potentia. Sickness of spirit is the sickness of those who have cut the ties with the transcendent part of the universe, who are not in harmony with its principles. *To have peace of mind means to live in accordance with the principles of mind—that is, the principles of transcendent physical reality.*

# Part 1

## In Search of
## the Transcendental Elements
## of Human Knowledge

*The Non-Rational and Non-Empirical
Elements in Rational/Empirical Knowledge*

# Chapter 1

# THE COMPOSITE NATURE OF KNOWLEDGE

## *The Conceptual Foundations of Science*

Science is central. By *science* I do not mean technique, technology, the basis of our survival; rather, I mean our view of the world, our quest for understanding the nature of reality, the order of the universe.

In the history of the West the nature of reality has always been taken as a central challenge. What are things really like? What is the position of human beings in the universe? What in our knowledge is certain and what is illusion? Is the external world independent of perception? Is it made up of material things or non-material ideas? Do the qualities of things belong to them or are they creations of our senses? Does eternal truth reside in universal principles or in particular events? In immutable forms or in evolution? Or is there no knowable truth and are the laws of nature made up by the human mind?

Questions of this kind are a challenge because they have never allowed for final answers. Like an open flame to moths, this challenge has been the fate of Western minds: enlightenment to some, annihilation to others.

Questions of this kind are central because, throughout our history, the answers that people have given to them have typically affected their views regarding the order of human concerns, political, social, and private.

**Socrates** (470–399 B.C.), for example, was fascinated by the faculty of reason to determine the *general features of things*. In accordance with this and, most likely inspired by it, he stressed the *authority of general principles* in the standards of human conduct.

In contrast to this, the *Skeptics* denied both, the possibility of *objective knowledge*, and the authority of *generally valid moral principles*.

**Kant** (1724–1804) emphasized the importance of *innate principles* for our theoretical knowledge and for our moral conduct. He claimed that the laws of physics were made by the human mind in the same manner in which, he claimed, the human will did not conform to external principles but to *categorical rules* set by itself.

These examples demonstrate that, regardless of what our convictions are, our views of reality effectively shape our moral convictions and form a basis for making decisions in daily life. By contemplating reality, we learn about principles of human conduct. *There can be no system of rules of human conduct (ethics) without a concomitant view of reality (ontology) or a theory of knowledge (epistemology).* In the order of the universe, human order is revealed. *"By gaining understanding of the world,"* **Aristotle** (384–322 B.C.) said, *"man comes to understand who he is."* That this should be so is perhaps a sign of deeper connections, of something in the background of physical reality that is akin to the human mind and communicates with it.

These connections between the descriptions of science and the prescriptions of ethics are somewhat hidden and frequently denied. It is typically claimed that the exploration of reality has nothing to do with human values. Science is supposed to be particularly incompetent in setting guidelines for human conduct, and to be indifferent to our hopes, fears, and general concerns.

Closer analysis will show that this view is not correct. For example, when our views of reality fundamentally affect our views of human conduct, one basic moral guideline immediately follows; namely, that it should be everybody's responsibility to maintain an enlightened and realistic view of the world. Anything else can entail dogmatic and unrealistic expectations which will disrupt society as much as the activities of a common thief. Other virtues, such as modesty, tolerance, and commitment to truth are also easily connected with epistemology.

*Naive realism* is the generally accepted view of science. It is the contention that the authority of science is restricted to *facts*, to *bits of knowledge established with incontestable certainty.*

# Establishing Facts

When people are asked to explain what they mean when they say *they know something for a fact*, they usually reply that they have something in mind, but can't say it. This is so because common sense is dogmatic, careless, and uncritically takes many things for granted. When pressed to say more about the basis on which they accept a thing as a fact, or what procedure they apply to establish a fact, most people produce some fragments, like a wish list, of necessary conditions and operations.

*Experience of a datum*—of something given, such as a reliable record on the past or direct evidence for a present fact—is a necessary part of verifying any fact. A fact is something that has independent existence, outside of anyone's mind, regardless of what anyone wishes, thinks, feels, or supports as a bias. That is, an existing fact can be *observed at will*, repeatedly, for the purpose of testing. Independence from individual observers is a necessary condition for *objectivity.*

*Compatibility with reason* is another essential aspect of the procedures that we use to establish facts. Verifications of facts must be reasonable. They will not be accepted if they make no sense. There are *laws of thinking correctly,* which are summarized in logic, the science of valid reasoning. If it does not conform to logic, it cannot be accepted as a verification of fact.

Factum, in Latin, means *something that has been done,* that is real, actual. *"As a matter of fact"* means *"really."* There is a connotation with *certainty, verity.* There is a feeling that if something is known for a fact, it is known for sure, with certainty, and that a proposition can be made about reality in which nothing is taken for granted.

*As it turns out, all propositions about reality and all techniques of establishing facts take a lot for granted, and the means of observation and reason that we employ in deriving facts are not as clearly and distinctly factual as the feeling of certainty that they evoke.*

For example, as to the desired *rationality* of established facts, the concept presupposes that the logic of the universe must in some ways conform to the logic of the human mind. Why that should be so is by no means self-evident.

Furthermore, as to the alleged *objectivity* of a fact, it is a complication that *stating a truth is a satisfying act,* with a feeling quite different from the sensation that we have in telling a lie. Moreover, discovering an unexpected fact is an exciting experience. Thus, the subject is emotionally involved in the process, and expressions of facts often are not purely rational and objective, but biased by emotions.

In addition, experience of reality and its memory are complicated processes. What exactly happens during an observation? What kind of contributions are made to the data by the system that records, stores, recalls, and communicates the memory content? In what way are the data biased by the censorship of the nervous system, which allows some of the sensory signals to proceed to consciousness while others are denied as insignificant? *No fact established by an intelligent mind exists by itself.* It is always a truth derived, attached to equipment, the result of a process, an interaction between an intelligence and an external phenomenon.

A digital computer, for example, receives numbers from its sensors and A/D (analog to digital) converters and stores them in memory. Analog signals are continuous, like a temperature, a frequency, a voltage. They can obtain any value by changing in arbitrarily small increments. In order for such an observable thing to be stored in memory, the measured results must be altered by A/D converters to discontinuous numerical values, or digits. *A temperature, or a pressure, or intensity of radiation, to a digital computer all are one: numbers.* Thus, when asked what the world consists of, a computer will answer "numbers." In the same way humans are led by their sensors to conclude that the world consists of objects hard and soft, hot and cold, heavy and light, dry and wet.

In this way *experimental data* in science are not like what has been defined here as facts. They are not truly given, but are a curious mixture of contributions from the external object, the detector, and the observing mind. In this sense, various traditions of Western philosophy agree: *We do not experience things, but only our interactions with things.* Whereas objective observation must be independent of the state, character, and nature of the observer, no observation can be so.

As we shall see in part II of this book, at the *elementary level* of matter, at the very foundations of observable reality, further complications arise. When dealing with microscopic systems, the means by which we observe—that is, the quantum or indivisible unit of energy—has properties of the same magnitude as the electron or atom with which it communicates. *Thus, observation unavoidably affects what it observes.* More complicated yet, in the world of the *quantum phenomena* many basic characteristics of things, dynamic variables such as position or momentum, are now believed by many not to exist if they are not observed; that is, they are *created by observation.*

## The Theorem of the Threefold Basis of Knowledge

We say of facts that they are *truths.* It is interesting that the English term *true* has the same root as the German word *treu,* meaning loyal and faithful. In a way, establishing facts has to do with faith. *There is something there you have to believe in.* One is reminded of **Saint Anselm** (1033–1109), who explained his faith by saying, "I believe to understand," and of **Pascal** (1623–1662), who said, *"The heart has many reasons that reason does not understand."*

Faith is essential to the process of deriving facts, because a number of *principles of inference* are involved which are *non-rational* and *non-empirical* in the sense that they themselves cannot be derived from reasoning nor established by observation. *Whereas the processes used in deriving facts must be rational and empirical, the principles used in these processes are not.* Among them we find the assumptions of *object permanence, induction,* and *causality.*

I.   The **principle of the continued existence and identity of things** is a basic assumption that we automatically apply in our observations of physical reality. Without assuming that all ordinary things have a continued existence and identity, coherent observations of the external world are not possible. However, uninterrupted observation of anything is impossible, for practical reasons in general and, as shown in part II, in particular because it violates Heisenberg's uncertainty principle. *In essence, object permanence is unobservable.*

II.   The **principle of induction** is an important principle of science. In *logic* induction is the procedure of moving from the particular to the general, of making inferences of a general nature on the basis of particular arguments. In *science* it is the process of formulating general laws on the basis of particular observations, assigning attributes which have consistently been found conjoined with *a few* events of a class to *all* events of that class. Induction in science is contingent upon the proposition that *the future resembles the past*, an assumption which cannot be derived by any process of reasoning nor verified by experience.

III.    The **principle of causality** is applied when the relation between two events is assumed to be one of cause and effect. If causality is a principle of nature, nothing happens without a cause. The principle is important in understanding physical reality, because meaningful observations of the external world are possible only because a signal causes a response, the response in turn may be another signal causing another response, and so on. There is a **causal chain** from perceiving to sorting, storing, and recalling a stimulus.

Yet, already in the eighteenth century **Hume** (1711–1776) argued that *we have no experience of any causal event. We always* **observe** *temporal conjunction, but* **infer** *necessary connection.* Thus, causality is not a principle of nature but a habit of the human mind. It is part of a *system program* of the mind whose origins have not been revealed to us.

According to Hume, elements of knowledge consist of *impressions*; the origin of impressions is in **sensation** and **reflection**. *Ideas* are copies of impressions. Ideas without impression are meaningless.

Principles for which we have no impressions include the *certainty of matters of fact* (the contrary is always possible); the certainty of the continued *identity of things* (only if causality holds); the validity of *induction* (the future need not resemble the past); *causality* (a habit of the human mind). Thus, the very principles needed for deriving knowledge in themselves are meaningless. Or, put another way, the principles we use in deriving knowledge cannot derive themselves.

In order to expand on Hume's arguments, consider the simple phenomenon of colliding billiard balls. When we propel a ball across the surface of a table, on impact with another it will suddenly change its direction of motion while the second ball will abruptly bounce off in a predictable way. In analyzing this phenomenon, we apply the principle of causality and we say that the first ball, by striking the second, *caused* it to move in a certain way. Why we apply this principle is not immediately obvious and not easily justified. Certainly we do not *observe* that the one ball causes the other to move. We merely observe that while one ball is changing its state of motion, a second makes a move on its own, in a certain way. *We observe temporal conjunction; but we **infer** causal connection and claim that one event was the cause of the other.* No otherwise established principle would be violated, no law of logic suspended and no experience contradicted, if our instructions would be to the contrary; that is, *be careful not to correlate two contiguous events, since nature is acausal.* However, since our instructions are *check the memory content for possible connection whenever there is temporal conjunction,* then this inference is made that one event is the cause and another, the effect, and the operational principle is faithfully executed like the sequence of statements in the subroutine of a computer program. Where the program comes from and why it should be true are important questions. *"If there is no causality,"* Eddington (1930) wrote, *"then there is no clear distinction anymore between the Natural and the Supernatural."*

By following various traditions of Western thought, then, we accept that the basis of human knowledge is threefold. The essential prerequisites are **experience of reality, employment of reason,** and **reliance on certain non-rational and non-empirical universal principles,** such as the certainty of matters of fact, the validity of induction, causality, lawfulness of nature, the continued identity of things. The latter represent a component in knowledge which is not derived by a *reasoning of the mind* (as a computer program is not derived from its hardware) *or by an experience of the external world* (as a computer program is not derived from its analog-to-digital converters). It is an element in our thinking that is beyond human control.

Where do the universal principles come from? Do they exist because, as **Hegel** has said, the real is rational and the rational is real? Is it, as **Plato** thought, that the soul remembers? Is it, as **Descartes** has said, because some things can be seen clearly and distinctly? Or is there, as **Leibniz** has said, preestablished harmony? In the face of all that has been set forth, one fact is unavoidable: there is an illegitimate element in our thinking that we cannot ascertain. There is a component in empirical knowledge that is derived neither from the mind a priori nor from experience of the external world. Rather, it is a component derived from an order or intelligence or level of information beyond human reason and experience. *Some unknown process has made it a part of the mind, and its inexplicable origin renders uncertain what seems most certain in our lives.*

In his book *The Science and Philosophy of the Organism* (1908) **Hans Driesch** (1867–1941) described the universal principles in our thinking—categorical principles in his terminology—in an excellent way:

> all categorical principles are brought to my consciousness by that fundamental event which is called experience, and therefore are not independent of it, but they are not inferences from experience, as are so-called empirical laws. We almost might say that we only have to be reminded of those principles by experience, and, indeed, we should not, I think, go very far wrong in saying that the Socratic doctrine that all knowledge is recollection, holds good as far as categories and categorical principles are in question.

## The Transformation of Hume's Problem by the Involvement of the Self-Conscious Mind

To rephrase Hume's problem: *The principles that we use to establish empirical knowledge cannot establish themselves.* All the processes that we employ in deriving knowledge about the world are based on principles which are not as certain as the products that they bring about.

Hume was right. There is no experience of necessary connection in any *external* phenomena. In contrast to this thesis, however, *when the self-conscious mind itself is directly involved in a causal event, as when its associated body takes part in a collision, or when the mind by its own free will is the cause of some action, then there is no doubt that causal connections exist.* Thus, it is not true that there is *no* experience of causal connections. Rather, experience of causality by the self is such that not only *can* some phenomena be causally connected; they *must be* connected so.

*I cannot be sure of the continued identity of any one thing. But I am sure of my own personal identity, even though I lose consciousness periodically. From my own identity I infer the identity of external things.*

*I cannot be sure of any effective cause in an external event, but if I am myself the cause, there is no doubt that causal relations exist. From this I conclude the existence of causes in other external events.*

*I cannot be sure that a collision causes an external object to move (I can detect temporal conjunction, never necessary connection), but if I am myself the target of the impact, then there is no doubt that an external agent caused the change in my motion.*

Thus the self-conscious mind has the important function of establishing the principles without which our existence would be impossible. It is by experience with its own self that the mind provides a basis for our convictions regarding the validity of empirical knowledge and the certainty of matters of fact.

**In summary:** the basis of knowledge is threefold, consisting of (a) the experience of reality, (b) the employment of reason, and (c) the application of universal principles (induction, identity, lawfulness of nature, causality). The process of deriving *rational knowledge* by *empirical methods* presupposes principles, or basic propositions about the world, that themselves cannot be derived from experience or by reasoning. The conclusion is that *the basis of science is unscientific.*

In order to further expand on these matters, related topics are presented in the appendices, including closer analyses of illegitimate components in human knowledge (Appendix 1); of the failure of intuition (Appendix 2); causality (Appendix 3); the relations between modern art and modern science (Appendix 4); and Popper's Logic of Science (Appendix 5).

## Transcendental Knowledge

Searching for transcendental elements in human knowledge, we found them in the universal principles by which we derive an understanding about physical reality. Having to do with knowledge, these principles can be adequately termed **epistemic.**

*Epistemic principles are transcendental because they are neither derived by a process of reasoning, nor by operations performed on physical reality. They are, simply, principles of the human mind. Thus, identity, object permanence, causality, external reality—all the requisites for a reasonable and enlightened life, albeit uncertain to experience and reason—are valid because they are transcendental principles provided by the human mind.*

By producing these principles, it is as though the mind remembered a higher order than can be found in the laws of logic or the visible patterns of physical reality. Thus, it is a valid question whether evidence can be found from physical science of this transcendent part of physical reality, where such a higher order might have its roots.

# Part 2

## In Search of Transcendental Physical Reality

*The Non-Material, Non-Real, Non-Local, and Mind-like Components of Physical Reality*

# Chapter 2

# THE WAVE-PARTICLE DUALITY

Search for transcendental reality is not a novel endeavor. It has been the task of religion throughout the ages. What is new in our age is that now the search can be undertaken within the domain of physical science, rather than outside of it or in opposition to it. Specifically, in this part we will explore to what extent evidence of the transcendent can be found in the world of quantum mechanics.

The starting point is the *wave-particle duality*. It is the characteristic of elementary physical entities—particles like electrons, protons, or atoms and molecules—to exist in states which evolve like waves when they are not observed and like particles when observed. The wave-like states are typically extended in space like any wave, but abruptly contract under observation to localized events or point-like particles. This phenomenon is, among others, at the basis of a new kind of mechanics—*Schrödinger's wave mechanics or quantum mechanics* (named after Austrian physicist **Erwin Schrödinger**, 1887–1961) in which the behavior of moving particles is not derived from calculations of trajectories of mass points, as in Newton's or classical mechanics, but from wave functions similar to those applied to lightwaves in optics.

The presentation below owes much of the framework of its ideas to the books by **Baggot** (1992), **Gribbin** (1984), **Heisenberg** (1952, 1962), **Herbert** (1988), **Polkinghorne** (1985), and **Stapp** (1993), which are strongly recommended as further reading material to obtain a more detailed picture than is needed for the purposes of this book. Additional details are also given in Appendices 6 to 16.

The real nature of the quantum wave functions which represent systems of mass particles in quantum mechanics is still a matter of dispute. As shown below, they provide *probabilities* for

27

possible outcomes of experimental measurements to occur. Since the probabilities are obtained by forming the square of the amplitudes of these waves, Heisenberg called them "probability amplitudes." The phenomena that they describe make it possible to conclude that *the basis of the material world is non-material; that the constituents of real things are not real in the same way as the things which they make; that reality is created by observation; that the nature of reality is both non-local and mind-like.*

## *Waves and Particles*

Everybody has some experience with **waves** and an understanding of wave phenomena. Most of us have seen water waves on a beach or sand waves carved by the wind on the surface of sand dunes. We have not *visibly* experienced sound waves, but can easily visualize such waves in air, periodic compressions and rarefactions of air molecules that strike our ear to create the sensation of sound.

All of the examples mentioned above are *matter waves,* wave phenomena that exist in some material medium. In a vacuum there are no sound waves. Without water there are no water waves. A second type of wave is represented by **light** and is different. Lightwaves are electromagnetic waves, oscillating electric fields correlated with oscillating magnetic fields, and they need no medium. They can travel through empty space (see Appendix 2).

In studying the world of quantum phenomena, yet a third type of wave was discovered: **quantum waves**.

Quantum waves are a third type because they are not only non-material—needing no material medium to propagate—but in addition **they are empty.** Lightwaves can travel in empty space, but they carry energy. Quantum waves also exist in empty space but carry no energy or any other mechanical quantity. They are simply numbers, numerical relations. Because they are empty, evidence of their existence is circumstantial; *we must think that the universe is a network of quantum waves because the observable order appears as a manifestation of their interference.* For example, molecules are interference patterns of the quantum waves of electronic atomic states. Bulk materials—gases, liquids, and solids—are interference patterns of the quantum waves of molecular states. *The reality of* *quantum waves is inferred from the expression of their interference in the observable patterns of reality.* To explain the meaning of these statements is the task of the following sections.

In Appendix 6 the physical properties of waves and particles are reviewed for those readers who are less familiar with such matters. From the material presented in Appendix 6 it is readily seen that typical wave phenomena include **diffraction**—that is, waves can bend around the corners of obstacles in their path—and **interference**—that is, waves can add in a characteristic way, forming interference patterns. When two waves of the same kind coming from different directions meet at the same spot in space, they superimpose, interpenetrate, interfere, building large amplitudes at some points by adding crests to crests, or cancel at other points by superimposing a crest with a trough. From this process, the interfering waves can reemerge as unharmed individuals, resuming travel along the original tracks, as though nothing happened.

In contrast, particles are different. They are discrete, non-continuous, individual lumps of matter, definite amounts of energy. They have mass, possibly electric charge, spin and momentum (defined as mass times velocity, see Appendix 6). They are subject to gravity and inertia. *They are hard, solid, filling space, exactly localized, isolated, precisely delimited.* Classical particles have a well-defined outline, a definite shape. When two particles meet at the same spot in space, they do not interpenetrate and superimpose, but collide, bounce off.

From this summary it appears that the properties of waves and particles are mutually exclusive. Waves interpenetrate, particles bounce off. Waves are continuous, particles discrete. Waves are extended in space, particles localized. Waves are non-material, particles massy. But never mind how mutually exclusive the properties of waves and particles, **quantum entities** (electromagnetic waves, protons, electrons, molecules) display them all. In some experiments these elementary entities behave like waves—forming interference patterns—while in other experiments they behave like particles—colliding, pushing, and bouncing around. In optics, for example, light follows the laws of wave propagation; in other experiments, like the Compton effect or photoelectric effect (see Appendix 6), it acts as though it consists of *light particles*, called *photons*.

The ability of elementary entities to display both the properties of waves and particles is the **wave-particle duality:** *Quantum entities observed act like particles; not observed, like waves.*

From this starting point, the description of a simple experiment, involving *the diffraction of electrons by an array of slits,* can be used to illustrate this strange duality, revealing essential secrets and unexpected aspects of the nature of reality.

## THE DIFFRACTION OF ELECTRONS BY SLITS

Three simple experiments demonstrate the extraordinary nature of quantum entities. The setup consists of a source that can emit either particles, like bullets, or electrons, or some sort of waves, like lightwaves. The instrument also contains a detector, **D,** sensitive to whatever is emitted. Between the detector and the source is a diaphragm, a screen with a number of slits (for simplicity, say two, **S1** and **S2**). They can be either open or shut.

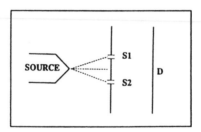

FIGURE 1
*Dual-Slit Apparatus for Diffraction Experiments*

**In the first experiment,** bursts of bullets emerge from the source and are sprayed across the slits without being aimed at any particular target. Some hit the screen, some fly through the holes directly, others ricochet from the edges. The detector records a hit each time a bullet arrives at it.

At first, let only the upper slit be open. In this case (1), after emitting bullets for some time, a bell-shaped pile of them will form behind the slit, indicating that most hits are recorded along the center of the slit, and fewer toward the edges. The bell-shaped curve of the pile can be thought of as a distribution curve of hits by bullets on the detector screen—we call it an **intensity distri-**

**bution curve.** If only the lower slit is open (2), the same kind of intensity distribution results, but centered on S2.

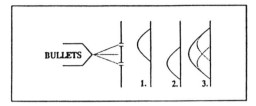

FIGURE 2

*The Accumulation of Ordinary Bullets Passing a System of Slits*
*(first experiment)*

In a variant of this experiment, both slits are open, each of them accumulating its pile. When enough bullets have passed the diaphragm, the two piles converge and their sum yields the total intensity curve. It is important to emphasize that, when bullets pass a system of several open slits, *total intensity is the simple sum of single-slit intensities.*

The experiment is entirely without any mystery. Bullets are ordinary classical particles. Each goes its own way and travels a definite trajectory from the source through a slit to the detector and lands as a hit, completely in agreement with the classical mechanics of moving particles.

If the bullets are emitted at random, it is impossible to predict where exactly at the detector a particular one of them will hit. When bursts of bullets are sprayed across a slit without aim, there is some probability for each of them to land in the center of the pile, and a lower probability to land on one of its edges. Thus the pile at the detector, or the intensity distribution, can be understood as a **probability profile:** with each bullet a certain probability can be associated to hit a given spot at the detector. The most important aspect, then, of the bullet-slit experiment is that **the probability profile from many open slits is simply the sum of the profiles from single slits.** This marks the end of the first experiment.

**In the second experiment,** the source emits some sort of waves, say lightwaves. They travel toward the screen, enter the slits, and hit the detector. Along its plane, the detector determines light intensity.

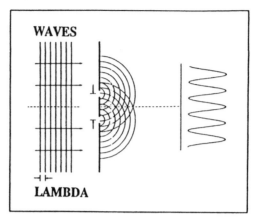

FIGURE 3
*The Diffraction of Lightwaves by Slits*
*(second experiment)*

If the size of the slits is similar to the wavelength (LAMBDA) of the light, the lightwaves passing through them are diffracted and a pattern of strips of varying brightness—a **diffraction pattern**—is seen on the screen. That is, each slit is the source of elementary wavelets which expand in all directions behind the slits, superpose, interpenetrate, and interfere. When both slits are open, the waves from one interfere with those from the other; their amplitudes add algebraically. In some areas on the detector, valleys of the wave coming from one hit on crests of waves coming from the other and both cancel. That particular spot is a dark spot. In other places, two crests are superposed and a wave with a large amplitude, or high intensity, results. Around this particular spot there is a region of brightness. (Intensity here is taken as a measure of brightness, synonymous with the number of photons hitting a given location on the detector plane.)

As in the first experiment, there is no mystery here. The diffraction of waves by a system of slits is a well-known phenomenon. From the wavelength of the light and the position of the slits, the pattern of intensities is easily explained. Like water waves on the surface of a pond, lightwaves interfere, they interpenetrate, and they superpose. Here elevated crests are formed, there the surface is smooth.

One aspect of this phenomenon is particularly important for understanding quantum waves. In the first experiment, when bullets piled up behind an open slit, two heaps from two slits simply merged to form one total pile. Total intensity was the *simple sum* of individual intensities. Each bullet traveled alone, found a hole, and delivered its impact to the detector, where it was recorded regardless of what the other bullets did.

Waves are not like that. When two slits are open, the intensity of interference at each point of the intensity pattern is the result of *interactions of waves that came through both slits at the same time*, forming a diffraction pattern that is typical for two slits. When the number of slits is changed, the pattern is changed. Close one of the slits, and the dual-slit pattern vanishes.

To understand the effect of superposition, it helps to consider that, in physics, the intensity of a wave is defined as the square of its amplitude (i.e., the square of the height of its crests). Following conventional notation, we represent wave amplitude by the symbol $\Psi$ (psi). In a two-slit system, when slit 1 is open and slit 2 is closed, light intensity is the squared amplitude of a single wave, say $|\Psi_1|^2$, where $\Psi_1$ is the amplitude of the wave coming from slit 1. Similarly, when slit 2 is open and 1 is closed, intensity is squared amplitude $|\Psi_2|^2$. When both slits are open, waves from slits 1 and 2 interfere and their amplitudes add, forming a single wave with amplitude $\Psi_1 + \Psi_2$. Intensity of the composite wave, then, is $|\Psi_1 + \Psi_2|^2$. This is different from the sum $|\Psi_1|^2 + |\Psi_2|^2$. *Thus, intensity of the composite wave is not the simple sum of the intensities of its component waves.*

From this it is easily understood that, if the number of open slits is increased to an arbitrary number, say *n*, an *n*-slit diffraction pattern will appear, with intensities that are the squares of the sums (superpositions) of *n* amplitudes, $|\Psi_1 + \Psi_2 + \ldots \Psi_n|^2$. Again, this is not the same as the sum of the *n* individual intensities, $|\Psi_1|^2 + |\Psi_2|^2 + \ldots |\Psi_n|^2$. Rather, each point of this intensity pattern is the result of superpositions of waves that passed at the same time through all the *n* slits. This is the reason why, in contrast to the merging piles of particles in the first experiment, a 2-slit intensity pattern of waves is not simply the sum of two single-slit patterns. When lightwaves are diffracted in the dual-slit apparatus, it may happen that a given spot on the detector will receive *less intensity when both slits are open than when one is closed*. This is the case when the crest of one wave hits the trough of another or if, at a given point $\Psi_1 = -\Psi_2 \neq 0$. In that case the sum of individual intensities, $|\Psi_1|^2 + |\Psi_2|^2$, would yield some finite brightness, but $|\Psi_1 + \Psi_2|^2$ will be zero. Thus, this spot gets **darker** when **more light** is allowed to strike. This could never happen when single heaps of bullets pile into one. When a single bullet piercing your heart will kill, two bullets passing through your heart will not ever cancel and keep you alive. In contrast, in the quantum world it makes sense to say, *Can we please turn some of the lights off, it is getting too dark here*. This marks the end of the description of the second experiment.

**In the third experiment,** the source is an electron gun from which electrons are emitted and projected through the slits and onto the detector. As we know, electrons are elementary particles, quantized lumps of matter with a definite amount of mass. They are too small to be seen directly by the human eye, but they can be detected by a fluorescent screen, among other devices. When an electron strikes such a screen a tiny flash is seen, a precisely localized event, a small dot of light emerging from a phosphor grain. Similarly, when detected by a Geiger counter, each individual electron can be counted as a single, isolated noise, a click; and when recorded photographically, each makes a tiny black dot on the photoplate.

All observations of electrons are of this kind. In each observation they reveal their corpuscular nature by a strictly isolated and localized event. When observed, they are tiny material particles, localized in space and impenetrable, point-like chunks of matter.

In the diffraction of electrons a large number is launched, one by one, through the apparatus. Each hits the detector and its impact is recorded. Like tiny bullets they leave the electron gun and a single flash is registered for each of them on the phosphor screen. Each single impact can be counted individually, as a flash or dot or click, like the impact of bullets in the first experiment. Little piles of dots accumulate behind the two slits. After a while we inspect them, and what do we find? Do we find a single heap behind each slit, as in the bullet experiment? Two piles, one behind each slit? Have two growing mounds coalesced, adding up to a single one? The stunning answer is *no*. For these little bullets, the electrons, instead of two merging piles we find a series of fringes of different intensities, a *diffraction pattern*, identical in essence with the interference pattern of lightwaves observed in the second experiment.

## THE ELECTRON PARADOX

There is a little *thing* here, an electron, that is always *observed* with the attributes of a material particle: it is a lump of matter, impenetrable, localized. As a particle it will get involved in the typical activities of its species, pushing others, colliding, dancing around like a billiard ball. When it passes through a diaphragm and we wait for it at one of the slits, it is always found in just one of them, like a regular bullet, never in several at once. When observed it is always a dot, a point, localized, indivisible, a tiny flash on a screen.

When not observed, *when it is not known* through which of the slits it is traveling, this so-called thing engenders a pattern that only the interference of waves can produce, that only the simultaneous passage of wavelets through *all* the open slits can bring about. When not observed, its behavior is that of a delocalized wave that can split into arbitrarily many elementary wavelets, self-interfering, taking *each* alternative path that is open to it.

Electron diffraction patterns cannot be formed by isolated mass points tracking along classical trajectories through single slits. To form such a pattern, some entity, some signal must have split up and traveled through all of the openings at once, penetrating twenty of them if twenty are open, interfering with itself in twenty different elementary components to which it has given birth.

*By this simple phenomenon, without the involvement of any mathematical analysis, without the mechanics of a theory that may be obsolete tomorrow, the amazing character of the basis of our world is being revealed.* **To summarize: Elementary physical systems evolve in states that are wave-like when they are not observed, but collapse to particle states when they are observed. While the wave-like** *states may extend through vast regions of the space-time continuum, the particle states are typically contracted to a small, localized spot.* When the electron is a particle, it is localized in space, cannot be split into smaller parts, retains its identity in collisions with others. When it is a wave, it is spread out over extended regions of space, merging with others when it interferes.

## CROWD WAVES

In the early days of quantum physics various arguments were put forth to reconcile such incompatible combinations of properties with normal views of reality. Among them, described by Herbert (1988), the concept of **crowd waves.**

We are familiar with water waves and are not in the slightest disturbed that they exist in a medium, water, that consists of discrete particles, molecules. Similarly, sound waves exist in a medium, such as air, that consists of discrete units, molecules of nitrogen and oxygen and others. Both water waves and sound waves are examples of *crowd waves*—wave-like phenomena that corpuscular media containing large numbers of particles can sustain. In the same way, the argument went, electron waves should be explained as crowd waves. Somehow, when a large number of electrons are crowded through a system of slits, they interfere and form a diffraction pattern, making waves as water molecules make waves.

Crowd waves can be uncovered by a simple test which Herbert (1988) called the **dilution test.** Put a bell under a bell jar and let it ring. Its sound will be muffled by the jar, but is audible. Now connect the jar to a vacuum pump and evacuate it. At one point, as the air in the jar is enough diluted by the pumping, the sound will break down, even though the bell keeps ringing. Sound waves are crowd waves; a high density of molecules is needed to support them. Whatever makes the sound pushes the air molecules together in periodic waves and the compressions and concomitant rarefactions set up a vibration in space. In the same way, electron interference phenomena were initially thought to be the property of *intense* electron beams, in which many electrons were crowded together. To reveal their nature as crowd waves, dilution tests were performed.

Experiments of this kind involve weak electron beams in which **single electrons** emerge at the source and pass through the apparatus, through the slits, and on to the detector. Each electron completes its path on its own, is alone in this setup and, when it arrives at the target, delivers a single flash on a phosphor screen or leaves a tiny black dot on a photographic plate.

Thus, one lone electron after another is propelled through the apparatus and its impact recorded. Each time, the release of a new projectile is held back until the previous one has met its fate and is gone, no longer available for interacting with others. In this way, as in the bullet experiment, one tiny black mark after another accumulates on a photo-plate. After some time, when the plate has been exposed to a large number of electrons, an intensity pattern appears, a density distribution, and we can now ask what kind of pattern it will be. Will it be like the piles of bullets, individual heaps of particles behind each slit, or will it be like the interference patterns, which are compatible only with the passage of waves through all the openings of the diaphragm at once?

*The surprising answer is that single electrons passing in isolation through a system of slits engender an interference pattern.* Electron waves are not crowd waves. Rather, it is as though **a single electron can interfere with itself**, acting like something that is extended through the entire apparatus and dividing into numerous wavelets.

The paradox is deepened by the fact that, as we have seen, wave properties are so discordant with particle properties. Waves can spread over extended areas in space, particles are confined to a single spot. A given wave can split into infinitely many elementary wavelets, that travel in different directions, while particles are indivisible and confined to a single trajectory. Waves can interpenetrate and emerge from this interaction unharmed, while particles crash together with other particles. Heisenberg (1962) put the dilemma in the clearest terms. "A certain thing," he said unequivocally, "cannot at the same time be a particle and a wave."

# The Conceptual Consequences

## PROBABILITY AMPLITUDES: THE BASIS OF
## THE MATERIAL WORLD IS NON-MATERIAL

There is a generally accepted principle in quantum mechanics according to which it is not meaningful to make statements about elementary particles when they are not observed. In the diffraction of electrons by an array of slits, at one time the emission of an electron will be registered at the source; at a later time its impact will be recorded at the detector. *At one time here, then there.* What the electron did in between these observations, or that it did anything in between, cannot be known and nothing should be said about it. As a consequence, exactly where a given electron will strike the detector is unpredictable and only the *probability* can be given that it will hit a specific point. Once emitted by the source, it has *some chance*—some probability—to be observed at a given coordinate of the detector, and *some chance* of being observed somewhere else.

Heisenberg (1962):

> Quite generally there is no way of describing what happens between two consecutive observations. It is of course tempting to say that the electron must have been somewhere between the two observations and that therefore the electron must have described some kind of path or orbit even if it may be impossible to know which path. This would be a reasonable argument in classical physics. But in quantum theory it would be a misuse of the language which cannot be justified.

In our discussion of the dual-slit experiments (see above), it was suggested that the bell-shaped intensity distribution of hits by bullets on the detector can be regarded as a probability profile. For the same reasons, an electron diffraction pattern can be interpreted as a probability pattern. Where the intensity of the pattern is high, the probability for an electron to strike is high; where the intensity is low, the probability to find electrons is low. Since the diffraction pattern is, as we have seen, the result of the interference of some wave-like states, the conclusion is suggested that the nature of these states is such that they contain information on probabilities.

This point is of crucial importance and needs further elaboration: the intensities of an electron diffraction pattern provide probabilities for electrons to strike. They are equal to the squares of amplitudes of waves—the quantum waves—which are the manifestations of the wave-like states in which elementary physical systems evolve, when they are not observed. At a dark spot, squared quantum wave amplitude is zero, intensity is zero, no electron will strike, and probability of a hit is zero. At a bright spot, intensity is high, squared quantum wave amplitude is high, and the likelihood of a hit is high. We can never see the quantum waves, but the *observable behavior of elementary particles—in fact all visible physical order—is determined by their interference.*

In order to account for the properties of quantum entities— their evolution in wave-like states and the expression of probabilities by the squares of wave amplitudes—in 1926 Erwin Schrödinger developed a new kind of mechanics—quantum or wave mechanics—which is radically different from Newton's or classical mechanics. Additional details are given in Appendix 7.

In Schrödinger's wave mechanics all elementary particles, atoms, and molecules—in principle, the whole universe—are considered as waves, mathematically expressed as *wave functions* and symbolized by $\Psi$. For the specific state of a given physical system, the wave functions are obtained by solving a wave equation, called Schrödinger's equation, which is similar to the equation used in optics to describe the propagation of lightwaves. The meaning of the wave functions is such that the squares of their amplitudes— we also say the squared magnitudes $|\Psi|^2$—correspond to probabilities. In Schrödinger's atoms, for example, *the electrons outside the atomic nuclei are wave patterns*—standing waves that do not propagate in space but are tethered at the nuclei. This statement does not mean that an electron outside an atom is running around the nucleus in waving circles, as in an epicyclic motion of Ptolemaic planets. Rather, *for all practical purposes the electron has dissolved into a quantum wave—some probability information—a different state of being than that of ordinary things.*

In applying the theory to the electron diffraction experiments, it is postulated that the apparatus sets up a system of quantum waves—probability amplitudes—which interfere, forming a network of interference patterns. The interference contains the probabilities of all the possible outcomes of measurements. We say that $\Psi$ contains all the possibilities of the system that it represents. Thus, in this process one cannot assert that a given electron has passed through a particular slit, if the corresponding observation was not made. Rather, *if not observed, the particle acts like it takes all possible paths open to it.*

In every instance in which it was tested, the nature of the functions which represent physical systems in Schrödinger's mechanics has been confirmed to be such that squared magnitudes, $|\Psi|^2$, will yield probabilities for actual events, for possible outcomes of measurements of system properties. Probabilities are **dimensionless numbers.** Probability waves are **empty** in that they carry no energy or mass. Numerical relations are their exclusive contents. Nevertheless, *all elementary particles must behave in accordance with the rules of interference of the quantum waves. These waves dictate, as it were, what, in the universe, is physically possible, and what is not allowed.*

*At the foundation of reality, we find numerical relations—non-material principles—on which the order of the world is based. The basis of the material world is non-material.* Heisenberg (1962):

> Modern atomic theory is essentially different from that of antiquity in that it no longer allows any reinterpretation or elaboration to make it fit into a naive materialistic concept of the universe. For atoms are no longer material bodies in the proper sense of this word . . . The experiences of present-day physics show us that atoms do not exist as simple material objects.

## SUPERPOSITIONS OF STATES:
## THE CONSTITUENTS OF REAL THINGS ARE NOT
## AS REAL AS THE THINGS THAT THEY MAKE

In discussing the diffraction of waves by an array of slits, it was noted (see above) that the observed diffraction pattern is the result of a *superposition* of waves that passed at the same time through all the open slits. The choice of language was not accidental and leads naturally to the concept of the *superposition of states.*

As we have seen, electron diffraction patterns are formed by some signal that splits into elementary wavelets and travels through all of the open slits at once, interfering with itself in as many different wavelets as there are open slits. Consider that, in a dual-slit apparatus, the passing of an electron through one slit alone, say the first one, defines a *state*, $s_1$, of the electron, represented by the wave function $\Psi_1$. Similarly, for the second slit a second state, $s_2$, represented by $\Psi_2$. The requisite motion that will engender the interference pattern is a *superposition of states*, of $s_1$ and $s_2$, corresponding to a state of motion in which a single electron is in both tracks, at the same time, through the first and the second slit, entirely disregarding the fact that electrons cannot be split into arbitrary fragments and reassembled at will.

We have also noted above that the intensity distribution of a diffraction pattern depends on the number of open slits. Thus, when an electron passes through this device somehow it must *know*, as it were, how many slits are open and how many closed. Something must be spread out through the entire apparatus to explore the status of its parts and *collect the information* that determines the intensity pattern that is observed. This is in seeming contrast to the fact that, when the electron on its passage is being watched, it is always found in *just one* of the slits. When being observed it is always a particle and the interference pattern breaks down. When not observed, the interfering waves immediately evolve and disperse through large regions of space.

Ordinary, real objects have a definite position in space. They *own* their piece of reality, whether someone is watching them or not. Having a definite position means that one does not have a chance to be at different places at the same time, or seems to mean that. When an electron appears to be passing, at the same time, through all the open slits of a diffraction device, it surrenders the right of owning a definite position in space. *Acting like it is everywhere, it is nowhere.* From this consideration arose the notion that the position of an elementary particle is **not an intrinsic attribute**— not a property that it owns, but one *created by a measurement*.

Considering the situation of ordinary objects, their probabilities in space are very simple. If an object is, say, at point $a$, then the probability to find it at $a$ is equal to unity or 100 percent, and zero everywhere else in the universe. For ordinary objects the probability is always 100 percent somewhere and zero everywhere else. They cannot do as quantum entities do and have non-zero probabilities at many places at the same time. Right now I am at this desk. My probability in space is not partially here and partially there and down the hall and outside this building.

As with position in space, other evidence suggests that, in general, dynamic attributes of elementary particles—such as their linear momentum or angular momentum—are not intrinsic, but are created by an interaction with a measuring instrument. This is the meaning of the suggestion that *reality is created by observation.*

Ordinary, real things do not behave like this. Ordinary things **own** their dynamic attributes exclusively and innately. They do not share the ownership of basic properties, such as position, with some nosy measuring device. *A person entering a room furnished with several entrance doors **must** come through just one of them, **will not ever** enter through several doors at once, or pretend that it could be done.* In contrast to this, the electrons act as though they were taking all possible routes open to them.

No ordinary thing can behave like the particles it is made of. The components of real things are not real in the same way as the things that they compose. Heisenberg (1962):

> What *is* an elementary particle? . . . If one wants to give an accurate description of an elementary particle, the only thing which can be written down as description is a probability function. But then one sees that not even the quality of being belongs to what is described. It is a possibility for being or a tendency for being.

In general, when a quantum entity can exist in a number of individual states, say $\Psi_1$, $\Psi_2$, . . . $\Psi_n$, in which a given property, like energy, has different values, the wave function that represents the system can be written as a **superposition**—a linear sum—of the individual states. In a superposition of states, the wave function $\Psi$ is said to contain all possible states, and it is further implied that each of them has some chance—a finite probability—to be found when a measurement is made. *Being in a superposition of states implies that the system is not in a particular state and nothing can be known about the property that the states $\Psi_1$, $\Psi_2$, . . . $\Psi_n$ represent.* Being real in the ordinary sense means to be in some definite state. When not in a particular state, a thing is not real in the ordinary sense.

In the eighteenth century, **Bishop Berkeley** (1685–1763) put forth the thesis that *"to be is to be perceived," "esse est percipi."* In Berkeley's philosophy, all things exist because they are in someone's mind. When not in a mind, they vanish from reality. In his *Principles of Human Knowledge* (1710) and *Three Dialogues between Hylas and Philonous* (1713), Berkeley proposed that **only the mind and its ideas are real in the world.** There is no reality external to the mind.

According to this theory, the world is made up of ideas and minds. Matter does not exist. *"It is plain that the very notion of what is called matter or corporeal substance, involves a contradiction in it."* Corporeal objects are not material, but complex ensembles of sensations, ideas in someone's mind. They exist by being perceived. If not in a mind, they cease to exist.

> Real things are ideas imprinted on the senses by the author of nature . . . The table I write on, I say, exists, that is, I see and feel it; and if I were out of my study I should say it existed, meaning thereby that if I was in my study I might perceive it, or that some other spirit actually does perceive it. There was an odour, that is, it was smelled; there was a sound, that is to say, it was heard; a colour or figure, and it was perceived by sight or touch. This is all that I can understand by these and the like expressions. For as to what is said of the absolute existence of unthinking things without any relation to their being perceived, that seems perfectly unintelligible. Their **esse is percipi,** nor is it possible they should have any existence, out of the minds of thinking things which perceive them. (Berkeley, 1710, 1713)

In Schrödinger's wave mechanics, the dynamic attributes of a system are expressed by an operator. A system property, say energy, is called forth—*evoked*—by *operating* on $\Psi$. That is, it is not really a property of $\Psi$. Similarly, a dynamical property of a quantum entity is not owned by it, but emerges in the act of a measurement. An observable attribute is not so much an attribute of a system, but a quantitative reaction to an operation performed on it. In this sense dynamic attributes are creatures of operations, while the quantum waves themselves are empty, carrying no energy. In this remarkable way, quantum phenomena have revived interest in Berkeley's curious principle.

*The unexamined life,* Socrates said, *is not worth living.* No wonder, since unexamined—unobserved, that is—it is not quite real.

## ARISTOTLE'S POTENTIA

According to views developed by Dirac, Heisenberg, von Neumann, and others, *reality is formed by two processes which reflect the two sides of the wave-particle duality.* Non-mathematical descriptions have been given by various authors, including Penrose (1989) and Stapp (1993). I follow Stapp's (1993) description and adopt his nomenclature in many instances.

**In the first process,** a physical system constantly evolves into a superposition of possibilities or tendencies, as Heisenberg called them, for actual events to occur. This is the wave-like state of reality, which Heisenberg (1962) called a *"probability function."* Stapp (1993) generalized the concept by calling it the *"Heisenberg state of the universe."* **In the second process,** the transition from the *"possible"* to the *"actual"* takes place—as Heisenberg called it (1962)—when an observation is made, in which case one of the states superimposed in the probability function is selected and becomes real in the ordinary sense. Such an event corresponds to the particle-like aspect of the wave-particle duality. Stapp (1993) called it a *"Heisenberg event"* or an *"actual event."* His description of the matter is particularly clear:

> Heisenberg's ontology has two elements, one of which accounts for the wave-like aspects of nature, and the other of which accounts for the particle-like aspects. . . . In Heisenberg's picture . . . the classical world of material particles, evolving in accordance with local deterministic mathematical laws, is replaced by the Heisenberg state of the universe. This state can be pictured as a complicated wave, which, like its classical counterpart, evolves in accordance with local deterministic laws of motion. However, this Heisenberg state represents not the actual physical universe itself, in the normal sense, but merely a set of "objective tendencies" or "propensities," connected to an impending **actual event.** The connection is this: for each of the alternative possible forms that this impending event might take, the Heisenberg state specifies a

*44*

propensity, or tendency, for the event to take that form. The choice between these alternative possible forms is asserted to be governed by "pure chance," weighted by these propensities. The actual event is simply an abrupt change in the Heisenberg state . . . The new state describes the tendencies associated with the **next** actual event. This leads to an alternating succession of states and events, in which the state at each stage describes the propensities associated with the event that follows it. In this way the universe becomes controlled in part by strictly deterministic mathematical laws, and in part by mathematically defined "pure chance."

The actual events become, in Heisenberg's ontology, the fundamental entities from which the evolving universe is built.

Because of its importance, here are the main features of this ontology again: between observations, physical systems evolve in wave-like states that represent tendencies for actual events. This process occurs in a continuous and deterministic fashion following Schrödinger's formalism, constantly dividing the wave function of the system into new branches—one for each possible outcome of a measurement—and constantly creating new superpositions of tendencies. The deterministic nature of the process is entirely analogous to the determinism of classical mechanical systems which obey Newton's equations of motion. If the Heisenberg state of a system at one time is fully defined in the Schrödinger formalism, details of its evolution can be extended into its future through any time span during which it is not perturbed by a measurement.

In contrast, when an observation is made, the wave-like state changes abruptly, discontinuously, and unpredictably in a *"quantum jump"* (Heisenberg, 1962). At our level of intelligence the process is seemingly ruled by nothing but chance, and a true *"choice"* is made.[1] From the many *possible* states superimposed in Ψ one will emerge in an observation as the *actual* state, but no predictions can be made as to which one will actually transform from the *"possible"* to the *"actual."*

---

1. Dirac quoted by Stapp (1993). The definition of *choice* given by Stapp: "any fixing of something that is left free by the laws of nature, as they are currently understood."

It is obvious that the wave-like Heisenberg state of a system is not an image of the reality that we know. As a superposition of tendencies of different unborn realities that have the potential to come into being, it has an entirely different nature than the realities that spring from it. Before the measurement, $\Psi$ contains all possibilities; thereafter, only the single, real one remains. Using a term introduced by von Neumann in 1932, we say that $\Psi$ *collapses* when it abruptly contracts to the form of the single state that a measurement made real. To the extent that a state is created in this process that did not exist before, we can say that elementary particles are not truly real when not observed and that reality is created by observation.

Heisenberg (1979):

> The indivisible elementary particle of modern physics possesses the quality of taking up space in no higher measure than other properties, say color and strength of material. In its essence it is not a material particle in space and time but, in a way, only a symbol on whose introduction the laws of nature assume an especially simple form.

and (1962, 1979):

> In the experiments about atomic events we have to do with things and facts, with phenomena that are just as real as any phenomena in daily life. But the atoms or the elementary particles themselves are not as real; they form a world of potentialities or possibilities rather than one of things or facts . . .
>
> The probability wave means a tendency for something. It is a quantitative version of the old concept of "potentia" in Aristotelian philosophy. It introduces something standing in the middle between the idea of an event and the actual event, a strange kind of physical reality just in the middle between possibility and reality.

Heisenberg's reference to Aristotle is remarkable. As a young man, Heisenberg discovered the similarities between the quantum phenomena and the teachings of the Greek philosophers. Specifically, Aristotle believed that **matter without form is not quite real.** He believed that stuff in itself, unformed and indefinite, is not part of reality, but it has the potential, *potentia*, to come into reality by being formed. *Form brings matter into reality.*

Things of the same kind may differ in many accidental features, but they share a common, unchanging, and necessary essence: **form**. Form can only be manifested if something is being formed. That *something* is substance, stuff, matter. Matter is the totally unformed and undetermined substance by which the forms appear. In terms of this concept, a Heisenberg event is the actualization, by an observation, of potentia. A Heisenberg state of a physical system is a state of potentia and involves the evolution of possibilities for future Heisenberg events. Each such event will put the system abruptly into a new state of potentia, which in turn will prepare the springing into being of other future events. After each collapse, the weaving of new patterns into new mixtures of potentia will immediately begin and continue, until another event disrupts the process. *This is how reality is created.*

**Potentia** is a concept in Aristotle's metaphysics that describes a state of being which, at the ground of reality, is intermediate between "not being" and "really being." In order for forms to become reality, matter has the meaning of possibility.

Aristotle's term for reality is *energeia*, the root of our "energy." His term for possibility is *dynamis*. It is an interesting coincidence that, in order for virtual particles to be real, they need *energy*, and that, furthermore, the values of *dynamical* variables of quantum particles are not quite real until they are observed.

It is not quite clear how Aristotle arrived at the view that forms help matter to become real, or how he was able to defend it. Sometimes it seems that inspired ideas need no defense and it is immaterial how they were derived. In any case, in Heisenberg's view the electrons and atoms share these aspects of potentia: when not observed, they are not real in the ordinary sense, but suspended in a world of possibilities, of non-material, wave-like patterns of probabilities. Particles in a superposition of states contain all kinds of wave forms; that is, they have no definite form. When a measurement is made, only one state—a definite wave form—remains. When measurements give particles a definite form, they project them into reality. In Aristotle's words: *Forms bring matter into reality.* In the terms of quantum physics: *Measurements give form (reality) to indefinite superpositions of states.*

## AT THE FOUNDATION OF LOCAL REALITY ARE NON-LOCAL, FASTER-THAN-LIGHT PHENOMENA

It is a part of Heisenberg's ontology that the probability functions represent *objectively existing tendencies, something that truly exists.* Since they are also typically extended in space, it follows that *Heisenberg events can be non-local events,* involving instantaneous changes of states across large parts of space.

Consider, for example, the case of an unbound electron, like one that was emitted by an atom after absorbing a photon. In between observations the probability function of such a particle will evolve in a state in which it will spread through extended regions of space. In the course of this process various possibilities can evolve to find this electron at various detectors located in different regions of the universe. The corresponding Heisenberg state will contain a branch for each of these possibilities. In the ensuing Heisenberg event (an observation or interaction of the electron with matter in a state of ordinary reality) the wave function of the electron will abruptly contract at an unpredictable but specific location to a single spot, or a localized event—producing, for example, a flash on a particular fluorescent screen, or a click in a particular Geiger counter. *At that same instant* the probabilities for this event at all the other detectors in the universe must drop to zero.

Similarly, in the electron diffraction experiments there is a finite, non-zero probability for an electron at one time to be found in every slit. Upon observation, at the spot of the impact the probability of its being there is 100 percent, and instantly zero elsewhere. *In this way the quantum processes contain an inherently non-local element, involving faster-than-light phenomena. As a result of something we do here now (making an observation) an instantaneous effect (a change in a local probability) occurs somewhere else.*

In the mechanistic world of material things—the world of classical physics—instantaneous actions at a distance are not allowed. Einstein's special relativity forbids the travel of any signal at a speed faster than light (see Appendix 2). The mechanistic foreground of reality is local, restricted to influences which need a finite time to travel from one spot to another. In a local universe, anything that happens in a distant galaxy will need at least as much time to affect us as a lightwave would need to travel here.

In contrast, if the universe is non-local, something that happens in its depths right now may have an immediate effect on earth. *In the Heisenberg ontology the nature of the quantum phenomena is intrinsically non-local. There are faster-than-light influences which are unattenuated by distance.* Recent experiments involving **Bell's inequality** have revealed additional evidence for the non-local nature of the universe. A description of the corresponding phenomena and of related topics, such as the **EPR paradox**, is given in Appendices 14 to 16.

## INFORMATION-SENSITIVE PHENOMENA: AT THE FOUNDATION OF ORDINARY THINGS ARE ENTITIES WITH MIND-LIKE PROPERTIES

In the electron diffraction experiments an interference pattern is formed only when nothing is known about the electron trajectories. If each electron's path through the array of slits is known, there is no superposition of states and no interference. Thus it seems that *a connection exists between the information one has on a given system and its observable behavior.*

As far as ordinary physical systems are concerned, such a connection is of course impossible. Ordinary objects are not affected by what one *knows* about them but by what one *does* to them; it takes a physical intrusion to change their macroscopic properties. For quantum systems, however, the situation is different. As it turns out, *quantum systems may respond in an observable way to changes in information, even when that information is obtained without physical intrusion.* The following example was taken from the excellent summary given by Horgan (1992). It represents an information-response experiment that was performed with photons by Mandel and his coworkers at the University of Rochester.

In the Rochester experiment the setup contained a laser, a beam splitter, two sets of mirrors, and down converters. Down converters are special crystals that actually split a single photon into two. A laser beam passing a down converter is thus split into two, descriptively called *"signal beam"* and *"idler beam,"* with *"signal photons"* and *"idler photons"* (Horgan, 1992).

In the Rochester experiment a laser beam was first divided into two beams, B1 and B2, by a beam splitter, BS, and the resulting beams were aligned by two mirrors, M1 and M2. The two beams then continued through two down converters, DC1 and DC2, with B1 passing through DC1, and B2 through DC2. At each of the two converters the photons were split into two, and two signal beams, S1 and S2, and two idler beams, I1 and I2, were formed. Thus, I1 and S1 were formed at DC1, I2 and S2 at DC2. Both idler beams were then combined and directed onto the idler detector, ID, while the signal beams were combined at the signal detector, SD.

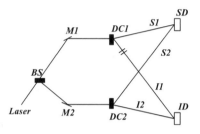

FIGURE 4

*Diagram of the Rochester experiment*

In this setup one cannot know which track a given photon will choose at BS. Thus, each single photon will travel through both converters, the two signal beams (S1 and S2) will be formed, and an interference pattern observed at SD. In this setup, Horgan (1992): "the signal detector cannot tell how the signals arrived." In contrast, when one of the idler beams is blocked, say I1, one can now deduce on which track an incoming photon traveled through the experiment. This is so because when I1 is blocked, the photons that arrive at ID must be photons of I2, originating at DC2. By comparing the arrival times of photons at SD and ID, one can then determine which way a recorded signal photon has taken. If the arrival time is the same, the two photons are twins, I2 and S2, that originated at DC2. In contrast, a photon arriving at SD without a twin at ID must have come via DC1.

When one of the idler beams is blocked off, we know which way a given signal photon traveled, and the interference pattern of the signal beams breaks down. The remarkable aspect here is that *it is not a physical disturbance of the signal beams, not some physical intrusion that caused this change in the outcome of the experiment. The inter-*

*ference involves only the signal photons. The intrusion involves only the idler photons. Signal photons and idler photons never meet again after they leave their down converters and yet, by blocking off the latter, the former are affected.* **The distinct change in a macroscopic observable phenomenon is seemingly brought about by nothing but the change in information about the system.**

Acting in response to changes in information is the prerogative of a mind. In this way, at the foundations of reality we find entities with mind-like properties, and a non-material, non-energetic principle—*information*—as an effective agent.

The mind-like properties of quantum particles are also suggested by numerous other phenomena. For example, one cannot help but think that, as probability waves, the quantum waves have a mind-like aspect. Probability functions are numerical relations, mathematical, mental entities. If non-material, non-energetic tendencies of this kind indeed have an objective reality, as assumed in the Heisenberg ontology, then they are something that is more mind-like than thing-like.

The mind-like properties of the background of reality are also suggested by the fact that its *order is determined by principles of symmetry,* abstract mathematical patterns, to which the constituents of the material world have to conform. Foremost among these principles is the **Pauli principle** (see Appendix 13): It is the basis for the existence of matter in the only form that is capable of supporting life as we know it.

At the level of elementary particles, idea-like states of being become matter-like; the Heisenberg tendencies are thought-like, and the results of Heisenberg events, matter-like. Actualization seems materialization. Greek mythology has often been invoked in this context. Whatever King Midas touched turned into gold. Whatever we touch by observing it turns into matter.

Most of the aggregates of matter can persist in stable mechanical states without in any way revealing their mind-like foundations. In human beings this is different: we are clusters of matter with its associated mechanical properties, but the idea-like state of being also breaks out at the level of consciousness. The mind-matter problem exists, because in its stable mechanical states matter forgets its origins in the world of idea-like states.

*Having* numerical properties or *having* associated probabilities and numerical relations is not what makes a state idea-like. But *being* an empty probability wave and *being* nothing but numerical relations is what turns the objectively existing tendencies of Heisenberg's ontology into thought-like entities. Somehow Ψ, in the middle "between the idea of a thing and a real thing" (Heisenberg, 1962), empty of matter and devoid of energy, seems to be closer to the world of ideas than to that of things.

Heisenberg (1962): *"The elementary particles in Plato's Timaeus are not substance but mathematical forms. 'All things are numbers' is a sentence attributed to Pythagoras. The only mathematical forms available at that time were such geometric forms as the regular solids or the triangles which form their surface. In modern quantum theory there can be no doubt that the elementary particles will finally also be mathematical forms, but of a much more complicated nature."*

## Transcendental Reality

In search of transcendental reality we found it in the world of the quantum phenomena. At the foundations of physical reality the nature of material things reveals itself as non-material; the elementary components of real things partake of a kind of reality that is different than that of the things that they form; local order is affected by non-local, faster-than-light phenomena; deterministic processes alternate with the expressions of choices in creating the visible order; and entities with mind-like properties are found. All these aspects are inferred from experience of physical reality, but they are beyond direct experience. They are at the foundation of the visible order of the universe, but they go beyond that order. They have all the characteristics of transcendent properties.

All this is in striking contrast to common sense and to the world of classical mechanics, in which all reality is reduced to the motion of particles which obey Newton's laws, a reality that has no room for the spiritual and mental, and does not even tolerate it because the nature of mind (non-material and not subject to determinism) is a violation of its premises. Mind is not a part or principle of Newton's laws and its willful actions are in conflict with those laws. In Newton's universe, mind is necessarily in conflict with matter. *But God, if God exists, must be a mind. In the Heisenberg events the word is becoming flesh—il verbo si é fatto carne.*

In the realm of the quantum phenomena, the mental enters the material world without any effort and interacts with it. Its forms of appearance are of a mathematical nature, typically assuming the guise of *information*. Interestingly, the term *"in-formation"* has the connotation of *"putting in a form,"* as in Aristotle's thesis that matter becomes real by being formed. To put something in a form is to in-form it. It is the form in material things that reflects what is real in them. It is the mind in us that creates a real person by "informing" it.

The electron diffraction experiments have opened a window to a different kind of reality, perhaps to the world of Plato's ideas, perhaps to a Divine Reality, where the mental may exist without a material substratum. The openness of the quantum world conveys a feeling of liberation, of being set free from the shackles of ordinary reality. *One might see the promise of messages—perhaps a benign guidance—from the depths of the universe, which affect our fate even though we do not understand them. And there is excitement in being part of a universe that is creative and where the unexpected and even the inexplicable constantly come into being.*

# Part 3

## In Search of
## Transcendental Human Nature

*The Non-Mechanical Aspects of
Living Machines*

# Human Mystery Versus Chance and Necessity

Searching for transcendent aspects of human existence, we have found them in our analyses of knowledge and at the foundations of physical reality. Knowledge extends as far as our sense experience and reason but it contains principles that experience and reason cannot verify. In probing the nature of physical reality, we find that the visible order of the universe is determined by stable material objects acting in accordance with local mechanistic laws, but the foundations are different than the ordinary human reality, revealing non-material, non-local, not-so-real, and mind-like properties.

If we have to admit transcendent elements in physical reality, we cannot deny them to human nature. However, since human nature is our own, it is the least understood of them all, and the existence or not of transcendency is a matter of passionate dispute, setting *mechanistic materialism* against the recognition of *mystery*.

The *human mystery*, as Eccles called it, is the existence, as the result of evolution, of human beings with values, of creative, inquisitive, and caring minds. In contrast is the thesis that all is the result not of mystery but of chance and the laws of physics and chemistry.

The question is whether *all the aspects of human nature are explicable in terms of the ordinary physical laws or not. Are additional factors at work, as yet not clearly identified?* In either case there is mystery, because *"explicable"* does not mean *"actually explained."* As it turns out, every time when new discoveries reveal hidden mechanisms of living organisms, new mysteries also arise.

What follows is a search for transcendent human reality using the arguments of two masterpieces, selected as brilliant representatives of contradictory traditions of thought. One of them is Eccles' book *The Human Mystery*. **John C. Eccles** was a physiologist and neurologist, born in 1903 in Australia. He has no problem in defending his position that there is more to human beings than mechanistic reductionism will explain. The opposition is represented, and not less passionately, in the book *Chance and Necessity*, by **Jacques Monod**, a French biochemist born in 1910.

# Chapter 3

# THE HUMAN MYSTERY

Eccles begins his discussion by asking whether, in contrast to the doctrine of mechanism, there can be a *natural theology* as a science like chemistry and physics, without reference to revelation? Can we, by observing reality, come to a conclusion about the existence of God? Do the facts of nature reveal a Divine Presence or Plan? **C. S. Sherrington**, one of Eccles' teachers, wrote about the same question: "The province of natural theology is surely to weigh from all the evidence derivable from Nature, whether Nature, taken all in all, signifies and implies the existence of what with reverence is called God; and if so, again with all reverence, what sort of God?" (Sherrington, 1940).

The human mystery according to Eccles is put in these terms: How has the materialist mechanism of biological evolution been able to bring forth beings with self-consciousness and human values? How can we explain the dualism of human nature—body and mind, and the interaction between mind and brain?

In physics, mechanistic materialism was the basis for explaining reality, but is no more. In biology, it is still used as a basis for explaining life. In the theory of evolution, it is the basis for explaining the development of beings with self-consciousness and values. In its most radical form—*monist materialism*—it is the doctrine that all aspects of humanity—our thoughts, memories, decisions, creativity in the arts and sciences, and values—are explicable in a deterministic and materialistic way, that no residue remains of a mental or non-material component.

To **Descartes**, animals and plants were machines. This is the biological equivalent of Newton's mechanism. In its modern form, the mechanistic view insists that, if a map of its genes is known, the organism can be constructed in every detail, including its hopes, fears, and dreams. In contrast to this, **vitalism** teaches that plants and animals are characterized by special agents—*life forces*—that act only in the biosphere. In taking an additional step **finalists** believe that some *design* was necessary for evolution to lead to self-conscious beings each with a unique individuality, and that natural selection is not accidental but guided toward some long-term goals.

**Polanyi**, the Hungarian biochemist and biologist, has expressed views close to vitalism in a modern way:

> All the processes in living cells are in agreement with chemistry and physics, but there is a hierarchic structure with emergence of higher levels that are not predictable from the properties of lower levels. The emergence of life is not predictable from the inorganic world; emergence of intelligence not predictable from the existence of algae . . . The operations of a higher level cannot be derived from the laws governing its isolated parts. That is, none of these biotic (of living things) operations can be accounted for by the laws of chemistry and physics. Yet it is taken for granted today among biologists that all manifestations of life can ultimately be explained by the laws governing inanimate matter. (Polanyi, 1966)

In opposition to monist materialism, various researchers, particularly Sherrington, Popper, and Eccles, proposed a *dualist-interactionist* view, according to which *the mind is as real as the material world, exists independently from the material world, and is able to interact with it*. In this view, mind and brain are independent entities which interact across a hypothetical frontier which permits, in a way not (yet) explained by science, the flow of information but not of energy.

The problem with the dualist-interactionist view as Sherrington (1940) saw it was that

> No attributes of energy seem findable in the process of mind. That absence hampers explanation of the tie between cerebral and mental. Where the brain correlates with mind, no microscopical, no physical, no chemical means detect any

radical difference between it and other nerve that does not correlate with mind. The two for all I can do remain refractorily apart. They seem to me disparate: not mutually convertible; untranslatable the one into the other. So our two concepts, space-time energy sensible, and insensible unextended mind, stand in some way coupled together, but theory has nothing to submit, as to how they can be so. Practical life assumes that they are so and on that assumption meets situation after situation; yet has no answer for the basal dilemma of how the two cohere.

We realize now that Sherrington's problem is a problem no more. In the quantum world, the mental and the cerebral—the mind-like and the matter-like—are no longer refractorily apart, but interact in an intimate way, matter seemingly springing from mind-like states. No flow of energy is needed in information-sensitive processes to affect the appearance of a macroscopic event. When physically energetic phenomena can be affected by the flow of information alone, it is not so unlikely to propose that the mind should be able to affect quantum systems in the same way, without the need for space-time energy sensible mechanisms. Alternatively, where interactions of mind and matter temporarily do violate the conservation of energy, Heisenberg's uncertainty principle (Appendix 11) offers a basis on which such processes might be allowed.

To Eccles monist materialism was an unacceptable doctrine because, as he put it, it was *"not a basis for a life with values."* There is no room for human minds in Newtonian clockwork. The mechanistic view of the universe destroyed any basis for the existence of God—"no need for that hypothesis," as **Laplace** peremptorily put it. In a mechanistic world, the feelings of freedom and responsibility are illusions and irritants. They are a part, nevertheless, of the great mystery in our existence.

The human mystery, then, is how to explain our own personal existence. *"This mystery,"* Eccles said, *"is the ultimate value in our world"* because he believed that we are creatures with a supernatural meaning, part of some great design.

## The Argument from Evolution

In his discussion Eccles (1979) presents a number of arguments against the mechanistic views of living organisms, including the argument from evolution. The evolutionary development of self-conscious beings was so unpredictable, this argument goes, and the course of evolution of human beings is such an unlikely event, that is is impossible to believe that this process was ruled by nothing but pure chance and the laws of mechanics.

To begin with, the process needed the creation and evolution of a universe and solar system whose properties must be exactly the way they are, or life as we know it could not exist. Change the fundamental constants of the universe in the slightest way, and our metabolism will break down. This is part of the *Anthropic Principle* (Barrow and Tipler, 1985).

Furthermore, Eccles points out, in our evolution *"dead-ends were avoided in a mysterious way, in that the species in our line of development were in some ways unfit, less efficient, less competitive, and therefore forced to continue on a narrow path that led to human beings."* The *fish* in our line of ancestry were *"poor swimmers"* who had no chance competing with the ocean racers, but their fins were adaptable as limbs for crawling. Conditions under water became rather desperate, and eventually these creatures were forced to crawl ashore and became amphibians. As amphibians they developed a remarkable *innovation*, inexplicably adding five digits to their limbs. Consider the importance of this peculiarity for the human species, and what the absence of this allegedly random acquisition would mean for the nature of human beings.

Next on our line of ancestry was a lowly reptile that barely survived its heavyweight competitors, the dinosaurs. Miraculously, the latter perished, and the first mammals developed on our line. Among these, the more *"impressive"* representatives—the carnivores, herbivores, whalefish—eventually all gave up their five digits. Instead, our line continued with the early primates, poor losers who had to live in trees because competition on the groundfloor was too fierce. Thus, they needed their fingers to hang on for dear life, and the significance of vision began to exceed that of smell.

The species leading to homo sapiens are characterized by their **plasticity**; they could be molded, were flexible, capable of undergoing significant changes. They were plastic at the cost of immediate success and the risk of extinction. *"A tenuous and unpredictable line of evolution led to us,"* an immensely improbable outcome in the game of evolution. The suggestion of guidance is hard to avoid.

"Does it not seem strange," Sherrington asks, "that an unreasoning planet, without set purpose and not knowing how to set about it, has done this thing?" (Sherrington, 1940)

"Biological evolution," says Eccles, "cannot account for the existence of each one of us as unique self-conscious beings." There were too many **"providential escapes"** in our development. Millions of years ago no biologist could have predicted that the human species—*"our particular gene combination"* as **Dobzhansky** (1967) called it—would eventually evolve.

At the stage of the homo sapiens a new process, cultural evolution, complemented and then replaced biological evolution. For the first time pressures for selection were derived from culture. Eighty thousand years ago the first evidence of self-consciousness and values appeared in ceremonial burial customs. There is evidence that, sixty thousand years ago, handicapped persons were treated with compassion as incapacitated Neanderthals were kept alive. This is the human mystery again: **"How are values derived in a process controlled by blind chance and survival of the fittest?"**

Culture is refinement and perfecting of the mind, of thought, emotions, manners, taste, brought about by study and training. Civilization means being socially organized. Ants can have a civilization, but not culture. Where did culture come from? The brain is genetically coded, cultural evolution is not. In the words of Dobzhansky (1967): *"Genes make the origins of culture possible, . . . but do not determine what particular culture develops, where, when, or how. Genes make human language possible, but they do not ordain what will be said. There is no gene for self-awareness, or for consciousness, for ego, or for mind."*

# The Argument from Popper's Three-Worlds Hypothesis

It was in opposition to doctrines of monist materialism that Popper developed the *three-worlds doctrine*, a tenet based on a special definition of reality, according to which something is real if it can affect the behavior of a large-scale object. Such real objects, substances, fields, all material things form **world-1**.

Brain states are real. They can affect the behavior of world-1 objects. But they are different from them and form **world-2**, the world of conscious experience—perceptual, visual, auditory, tactile, including pain, hunger, anger, joy, fear, memories, and thoughts.

Beyond world-1 and world-2, there is **world-3**, composed of the products of the human mind. It is the world of public knowledge, of all linguistic expressions, of enduring records of intellectual human achievement, libraries, museums, the world of culture. Paintings, ornaments, ceramics, tools, a world of storage. Much of world-3 consists of world-1 objects transformed by world-2. For example, a piece of music is not just a piece of paper and some ink used to write the score; it is not just a grammophone record, a CD, or a tape; it is not simply sound vibrations. These are all world-1 objects that exist in their specific form because of the music.

The piece of music is more than an object of world-1 and is real because its existence can affect the behavior of large-scale physical objects. But this is possible only because of the intervention of a conscious human mind. Without a mind world-3 is not real, but it has the *potential* of being real. Read this in light of quantum interpretations, according to which the results of measurements are real only when they are observed.

Extending this argument to the reality of the self-conscious mind itself, Popper and Eccles (1977) claimed that the mind is real, different from any physical object, and separate from body and brain. World-1 events can occur when the thoughts of the mind interact with the physical states of the brain. That is, at some point there are changes in the brain that do not result from energetic physical causes.

In world-1, invisible substances are real (e.g., air) because they can affect other visible objects. In world-2, the states of the brain are real. They send signals along nerve lines, cause muscles to contract, hit an undoubted world-1 tennis ball. World-3 entities are not just physical objects, nor brain states. Stories, myths, pieces of music, mathematical theorems, and scientific theories all need the intervention of a self-conscious human mind to become real.

According to Eccles: *"The human brain is open to influences of a non-material character from world-2. By the evolutionary process creatures had been developed with brains that allowed them to transcend the hitherto unchallenged world of matter-energy. There was no longer a complete closeness of world-1, and this momentous change has resulted in an on-going transformation of Planet Earth, as told in history. We have now come to an expression of the ultimate in human mystery."* **The parallel is exciting and inescapable. In the quantum phenomena the world of mass-energy is transcended. The universe is not closed anymore but, as in the dual-slit experiment, open to a different type of reality. The basis of the material world is not material.**

## The Argument from the Creation of a Self

In addition to the amazing **string of contingencies** that has led to the evolution of human beings, it is mysterious how mind and selfhood come to be from a single cell.

Sherrington (1940): *"The initial stage of each individual is a single cell. We have agreed that mind is not recognizable in any single cell we ever met. And who shall discover it in the little mulberry mass which for each of us is our all a little later than the one-cell stage; . . . yet who shall deny it in the child which in a few months' time that embryo becomes. Here again mind seems to emerge from no mind."*

A single fertilized ovum, the zygote, is responsible for building, by repeated mitosis, all the differentiated parts of the body, muscles, membranes, blood cells, nerve cells, the eye, the cochlea. Each cell of the body has been derived from the same single initial set of DNA molecules, from the same information, and yet, this tremendous differentiation takes place spontaneously.

*"To believe that this process,"* Rupert Sheldrake (1981) wrote, *"occurs without a design, without a masterplan, automatically, guided by nothing but the laws of physics and chemistry, is equivalent to the belief that, after [delivery of] a ton of bricks, cement, wood, and glass to a construction site, the new building will erect itself spontaneously."*

Among the many types of cells that emerge by differentiation from the singular source, nerve cells (*neurons*) are special. Once the primitive nerve cells are formed, their pyramidally shaped body (*soma*) never divides again, leading an independent biological life. They grow an *axon* that projects downward from the center of the soma, then *dendrites*. There are short spines on the dendrites and the axons develop fine branches which end in little knobs. The knobs of one cell can make contact with the surface of another and two otherwise isolated neurons communicate in this way. The contact areas were called *synapses* by Sherrington (Greek, *synapto*, to clasp tightly); a single neuron may have thousands of synapses on its surface. Transmission of information in the nervous system is by brief *electric impulses* along nerve fibers and across synapses. Impulses are generated by a neuron and discharged along its axon when it has been excited synaptically. There are *excitatory and inhibitory* synapses. The first cause the recipient neuron to fire an impulse down its axon, the latter inhibit a discharge. The brain consists of *two cerebral hemispheres*, linked by an organ called the *corpus callosum*. The external layers (*neocortex*) of the hemispheres consist of about 10,000 million neurons. The structure of the neocortex is further divided in quasi-discrete associations of neurons, called *modules*.

In addition to the development of a functioning brain, input from world-3 is necessary to develop a human self. Popper and Eccles (1977): *"As selves, as human beings, we are products of world-3, which, in its turn, is a product of countless human minds."* The self-conscious mind (world-2) evolves by continued interaction with world-3. The brain is necessary to the development of world-2, but not sufficient. The more world-3 resources an individual possesses, the more it gains in world-2. In a process that promotes all cultural evolution, world-3 helps a growing human being to achieve in a short time what biological evolution needed thousands of years to do. In an impoverished cultural environment, children mature to a limited cultural existence. In a violent cultural environment they mature to a violent existence.

Cultural evolution is different from biological evolution. Biological evolution was needed to develop brains. It is *genetically coded and inherited*. Cultural evolution is not. Each child builds its own culture from scratch. The nature of culture is that of an acquired character.

*"The brain in world-1 and the world of culture in world-3 are necessary for the development of the conscious self in world-2, but they are not sufficient.*

*Each of us knows the uniqueness of our personal self. The coming-to-be of each unique selfhood lies beyond the field of scientific enquiry. It is my thesis that we have to recognize the unique selfhood as being the result of a supernatural creation of what in the religious sense is called a soul.* "(Eccles, 1979)

In the evolution of world-3, permanent records play an essential role. About fifteen thousand years ago the cave paintings of Europe were among the first permanent pictorial creations. Around 2000 B.C., in the Mesopotamian river valleys, the Sumerians, under pressure to maintain a complicated administration, to preserve laws and traditions passed down orally, invented the written language; abstract letters, the **cuneiform**. This was one of the greatest achievements of human history. By it, for the first time man could live beyond time. First examples are business transactions, contracts, inventories, then royal inscriptions, then religious texts.

## The Argument from the Mystery of Conscious Perception

In perception, receptor organs first select, then encode environmental gradients into a discharge pattern of electric nerve impulses, expressing the intensity of a stimulus by the frequency of discharge.

Across synaptic linkages at *relay stations,* these signals from the receptors are transmitted to the higher levels of the central nervous system and ultimately to the levels of conscious vision, hearing, and touch. It is now believed that control and censorship of relay stations is used to sharpen the signal, eliminating weaker excitatory action, and suppressing those that at a certain moment seem irrelevant. For example, when a person is concentrating intensely on some matter, even severe signals will not be allowed to interfere. In general, it would be highly confusing to process all the data that flood the brain through millions and millions of afferent fibers. Some selection has to be made, some discrimination enforced, but it is not known how we make it or how it may bias the result.

## VISUAL PERCEPTION

In visual perception, the image is optically focused on the retina, a sheet of closely packed receptors, cones and rods. There the optical image is *fragmented* by millions of pixels—specialized structures—into as many independent signals, abstractions of brightness and contrasts of colors. The fragmentary codes arrive in the brain through independent channels. It is a complete mystery how the picture is resynthesized from the fragments to yield the subjective sensation of the visual wholeness of reality, creating conscious perception.

Eccles (1979): *"In some quite mysterious way, the retinal picture appears in conscious perception . . . a mysterious transformation occurs in the perceptual process . . . There is no colour as such in the brain . . . The human mystery is revealed by this extraordinary dichotomy between the coded performances of our cortical modules on the one hand and the perceptual experiences on the other."*

The picture that was focused on the retina, and its resynthesis, are nowhere found in the brain. You cannot open the skull and inspect it on a tiny monitor. All that is there are the coded forms of the impulse discharge patterns in the neural pathways. There is no color in the things, and none is found in any of the neurons, and yet color is in our perception. If it is not in the things and not in the brain, where is it? Where does the mysterious transformation occur in the perceptual process? Where does possibility become actuality? Where are the melodies heard after a sequence of sounds strike the ear? Where do serial noises turn into poetry? In short, where is reality created?

Eccles answers this question with **the dualist-interactionist hypothesis**, according to which the resynthesis of the visual image results when the self-conscious mind *"scans and reads out from the appropriate feature-recognition areas"* in the brain. In the dualist-interactionist hypothesis, *"mind and brain are independent entities,"* the self-conscious mind is an immaterial entity distinct from the brain. In the brain, the coded information from the receptors is contained in spatio-temporal impulse discharge patterns. From these codes, synthesis of the wholeness of experience is achieved by the self-conscious mind who is not a part of the brain. From outside of the brain, the independent self-conscious mind perceives the synthesized picture as it scans and reads the modules involved in the perception.

The dualist-interactionist hypothesis implies a brain-mind interaction through modules of neurons that are open to the mind. By direct action on open modules, the self-conscious mind can exercise indirect action on other parts of the brain. This interaction is not a process of energy transfer in the ordinary sense but involves the flow, across the mind/brain frontier, of information. In a similar way, in photon interference experiments, the mere flow of information can lead to a breakdown of the wave interference pattern, illustrating *that direct intrusive interference (energy transfer) is not needed to achieve energetic changes in a physical system.* There is no way to overstate the significance of this discovery.

Materialist Theories of Mind include (1) *Radical Materialism,* which holds that mental states do not exist; (2) *Panpsychism,* which holds that everything is mental and that the nature of all matter is rational; (3) *Epiphenomenalism,* which holds that mental states exist in connection with some material phenomena, but cannot affect any material structures; that is, world-1 is completely sealed; and (4) *Identity Theory,* which holds that mental states are special states of some material structures, such as the brain, and can affect material phenomena but that only the material structures are needed to engender mental states, so that the material events and the mental states are really identical.

In contrast to all these, the **Dualist-Interactionist Theory:** Brain is world-1; mind is world-2; both are separate and independent. Across a frontier, the *liaison brain,* there is a two-way flow of information. The world of mass-energy, world-1, is not completely sealed.

> The self-conscious mind is actively engaged in reading out from the multitude of active modules at the highest levels of the brain, namely in the liaison areas . . . The self-conscious mind selects from these modules according to attention, and from moment to moment integrates its selection to give unity to the most transient experience. Furthermore, the self-conscious mind acts upon these modules, modifying the dynamic spatio-temporal patterns of the neuronal events. Thus, the self-conscious mind exercises a superior interpretative and

controlling role upon the neuronal events both within the modules and between the modules.

A key component of the hypothesis is that the unity of conscious experience is provided by the self-conscious mind and not by the neuronal machinery of the liaison areas of the cerebral hemisphere . . . I conjecture that in the first place the *raison d'être* of the self-conscious mind is to give this unity of the self in all its conscious experiences and actions. (Eccles, 1979)

So far a view of the nature of humanity that admits great mystery—the transcendent beyond the laws of chemistry and physics. It is an attractive position to many, unavoidable to some, unacceptable to others. Among the opponents is Jacques Monod, whose arguments are summarized in the following chapter as found in his book *Chance and Necessity*. For Monod, not only is there no mysterious driving force for evolution and no masterplan for the fate of humankind, but in addition the craving for any principles other than blind chance and necessity are to him signs of the *"sickness of the modern spirit,"* and of *"man's primitive, animist origins."*

# Chapter 4

# CHANCE AND NECESSITY

## Tradition of Objectivity Versus Tradition of Purpose

**Vitalism** is a collective term for all theories that assume the existence of a special *life force, vis vitalis—a teleonomic (purposeful) principle,* Monod calls it—operating only in living organisms.

**Animism** is the belief that every object, animate or not, has a soul and is, in a way, alive. It is the assumption of a *universal teleonomic principle* that is active throughout the entire cosmos. *"The animist belief,"* Monod (1971) writes, *"consists essentially in a projection into inanimate nature of man's awareness of the intensely teleonomic functioning of his own central nervous system. It is, in other words, the hypothesis that natural phenomena can and must be explained in the same manner, by the same laws, as subjective human activity, conscious and purposive."* The important function of animism is that it conveys to human beings the feeling of a kinship with nature, providing a *"covenant between nature and man, a profound alliance outside which there seems to stretch only terrifying solitude."* However, at the basis of the covenant, Monod believed, is a fundamental error, the *"anthropocentric illusion."*

Among our traditions of thought, the tradition of modern science is based on *objectivity,* the belief that scientific description of nature must be independent of the wishes, moods, and needs of the individual scientist and that nature itself is objective because it is ruled by laws and not by concerns or intentions for its inhabitants. The laws of nature are inviolable, quantitative, inexorable, and general; they do not bend for a purpose. They apply regardless of the needs of a thing, exclude the rule of purpose in mechanical processes, and make no exceptions for mitigating circumstances. This is the tradition of objectivity in science. It is in direct contrast to all traditions of purpose.

When a process is driven by its end or occurs for a reason, it is said to be the expression of a *final cause* (Latin, *finis*, end). The purpose of an action is everything that one tries to achieve by it—the goal of the action, its end, its cause. Mechanical processes are aimless, purposeless; they follow the laws of motion, cannot have a goal. The driving agents of mechanical processes are *efficient causes*, and only the relationship between a cause and its effect—with no regard for any other considerations—is admitted to explain what is going on.

The quantitative natural sciences deal exclusively with the study of efficient causes and of the laws that they express. *Teleology*—the science of the purposes and final causes in nature—is in conflict with the methodology of modern science because the conception that physical phenomena are manifestations of purposive relations contained in them leads to value judgments in scientific arguments which are (1) **non-quantifiable**, (2) **non-testable**, and (3) **not expressions of general laws**. All this is in addition to the fact that the search for purpose is an irritant in science because it ranked highly in Aristotle's philosophy, and scientists have not forgotten that, at the end of the Middle Ages, Aristotle's science was part of a dogma that was enforced with violence.

Concerning a free-falling object Aristotle asked the question, *why does the stone drop to the floor?* His answer was, *because this is where it belongs.* According to the teleological view, objects enter into a state of motion *in order* to reach the place in the universe where they are meant to be. The greater their desire to get there, the faster they move. In contrast, modern science will ask, *how does an object fall?* This question asks for quantitative measurements, for experimental tests; it is the expression of a search for general laws that enable quantitative predictions. It is a productive approach, potentially leading to new discoveries and advances in knowledge. In contrast, the teleological view is in possession of the answers before any observations are made.

Leaning on a statement by Hume and modifying it, it can be said that no object or event ever discovers, by the qualities which appear to the senses, the future goal that it pursues. Teleological statements cannot be tested by experiment because they refer either to unobservable future events or to inherently untestable principles. Who would want to test the postulate that heavenly objects move in circles for the serene sake of not getting anywhere; that objects fall to the floor because they belong there; or that the processes in the universe occur for the sake of fulfilling a Grand Design? Furthermore, in a teleological process the ends—not some laws—justify the means, and objectivity is lost when the actions of every object are justified by its private purpose.

In considering the nature of living organisms in the framework of these two mutually exclusive traditions of thinking—of objectivity and of purpose—Monod described a peculiar paradox that is encountered in the biosphere. *On the one hand, scientific studies of living organisms, like of everything else, must be objective. On the other hand, objective studies of living organisms reveal that their essential properties include principles related to purpose.* Among these, Monod lists **teleonomy, autonomous morphogenesis,** and **reproductive invariance.**

*Teleonomy* denotes the state of having an ultimate purpose (Greek, *telos,* end; *nomos,* law). The paradox is that, even though all living beings are ruled by objective laws in every aspect of their existence, they are also "objects endowed with a purpose" (Monod, 1971). This is as far as, for example, the use of organs is concerned, or the purpose of the proteins synthesized by the body. The eye, for example, "represents the materialization of a purpose." *Autonomous morphogenesis* denotes the phenomenon that living beings are *"self-constructing machines."* In addition they are self-reproducing and reproduction is *invariant.*

The paradox shows that, even when vitalism is systematically rejected, one may nevertheless be forced to accept some of the same principles that vitalists employ. The processes of biochemistry are manifestations of objective laws, but the living beings engendered by these processes are manifestations of purpose. Because of the clear emphasis that is given in his book, I will call this **Monod's paradox,** even though related ideas have been expressed by various other authors.

Monod's paradox—*scientific description of nature must exclude any reference to final causes, but must admit teleonomic structures and purposive processes in the biosphere*—signifies an **epistemological contradiction** and is of great significance. Like the wave-particle duality in the realm of physical phenomena, this alien element in ordinary physical reality—teleonomy—is perhaps the sign of something deeper, of an order that goes beyond the mechanical lawfulness of living machines.

## The Laws of Physics and Chemistry

The functions of proteins—long chains of amino acids—are most important for our biochemistry. In the cells of living organisms about twenty different amino acids that occur naturally are combined to synthesize millions of different protein molecules.

The chemical linkage of amino acids in proteins is the *peptide bond;* proteins are *polypeptides.* By forming a peptide bond, each amino acid links its head to the tail of a preceding acid and its tail to the head of the next one in line in the peptide chain. In this process the free acid surrenders its independent existence and becomes what is generally called an amino acid *residue.*

Many proteins are chains with hundreds of residues. Residue sequence along the chain is determined by the genes of an organism, the *DNA content* of the cells. Genes control what proteins are synthesized in a living cell. When the chain of a given protein has been formed, in many cases it will *fold* onto itself in a characteristic way, forming the *single structure* in which it is active. In the most abundant proteins the structure thus formed resembles a globe. Hence the name **globular proteins**. Protein activity involves **catalytic functions** (helping chemical reactions to occur), **regulatory functions** (controlling what reactions in a cell will occur to what extent), and **constructive functions** (providing the membranes, organs, and structural materials supporting a body).

*Thus, without any mystery: genes determine residue sequence, sequence determines protein folding, folding determines structure, structure determines function, function engenders life.*

The general law of the sequences in naturally occurring proteins is that of randomness. When the first complete sequence of a globular protein was determined by **Sanger** in 1952, the resulting structure was, as Monod put it, *"both a revelation and a disappointment"*: there was no regularity, no special feature, seemingly nothing but random linkages of various amino acids. If all but one residue in a given protein are known, nothing in the existing sequence will allow one to predict precisely which amino acid will form the missing link. If a number of amino acids were shuffled together like a deck of cards to form an artificial sequence, the result would be indistinguishable from a sequence found in nature. However, once a particular sequence has been genetically coded, it is absolutely **invariant**, faithfully repeated, *never* synthesized at random, and its order is fixed, even though it is *randomness reproduced over and over.* This is one part of the story of chance and necessity.

Another part concerns the passing on of hereditary characters from parents to offspring. The replication of genes—the DNA molecules received from the parents, carrying the information needed to form a new being—is an invariant process and ruled by the *laws of stereochemistry.* Each new set of genes is a faithful copy of the original. It involves the separation of two intertwined complementary strands of DNA which later recombine spontaneously.

In Plato's world, all ordinary things were imperfect copies of perfect ideas. In the same way, the *idea* of faithful DNA replication is absolute, but the *real process* is not. It is a molecular process and as such is subject to quantum uncertainty. Mutations—errors—due to random perturbations can occur. Their accumulation in individuals leads to *aging and death;* in a species, to *change and evolution.* Since the mutations are seemingly caused by nothing but chance, Monod believed **chance to be the sole source of innovation.**

There is complete independence between the occurrences that can provoke or permit an error in the replication of the genetic message and its functional consequences. The functional effect depends on the structure, on the actual role of the modified protein, on the interactions it ensures, on the reactions it catalyzes—all things which have nothing to do with the mutational event itself nor with its immediate or remote causes, regardless of the nature, whether deterministic or not, of those causes. (Monod, 1971)

Once an error has been made and the mutation put in place, it is **rigorously and invariantly** incorporated in the system, and from then on is faithfully reproduced by the organism until the next error occurs. *Chance has become necessity.* The congruence is striking between this and the processes by which physical reality is created: when they are not perturbed, the genes are reproduced in a deterministic process, like quantum entities evolve in a deterministic state when they are not perturbed. When perturbed, a choice is expressed in the one case as in the other.

It is worth pausing at this point to consider the immense importance for the biosphere of the discovery that the nature of the universe is non-local. At the beginning of our existence we are but a single set of molecules. Their replication is a quantum process, involving the transition from a superposition of states—perhaps a number of characters that *might* be our own—to one definite, final state. In its most vulnerable, the most receptive stage of existence, what kind of messages might there be for this organism that comes into being, from events that occur in the depths of the universe and are connected by Bell non-locality? Not messages of the kind that our conscious mind would understand, but changes in phase relations that are clear to a system in the Heisenberg state. The non-local effects are random to human understanding but, once incorporated into the genes, they are pieces of information whose meaning will be expressed by the developing life. Similarly, it is impossible to say what effects the intrinsic quantum randomness might exert on this precarious emergence of an individual from potentiality, or what the fleeting fields of virtual particles might accomplish as they pop in and out of nothing.

One of the characteristic polarities of Western philosophy involves Parmenides' and Plato's views that *there can be no change, there is no becoming, eternal truth resides in immutable ideas,* and Heraclitus' thesis that the *nature of reality is based on flux and evolution.* These antipodal traditions have left their mark on later generations which have adopted either one or the other. As far as the living organisms are concerned, they have adopted both—**chance and necessity, invariance and evolution**.

There is a constantly recurrent motif parallel to the question of physical reality: as there is chance and necessity, there is causality and free will; the immutable idea—DNA—and the imperfect processes of material reality; the Heisenberg state of the universe and Heisenberg events—the deterministic evolution of not-quite-real tendencies, and the expressions of choices in creating true reality. One cannot help but speculate that the congruence is the sign of a deeper connection.

## Against Eccles' Argument from Evolution

In a large population of a species mutations must occur. It is not the evolution of species, but their stability that is amazing. In that process, Monod thought, *"blind chance can lead to anything, even vision."*

To Monod the evolution of life is the result of a *gigantic lottery*. Its irreversibility is simply the expression of the second law of thermodynamics in biology. Pure chance, no master plan, led to the formation of humanity. Its discovery *"is the last fatal blow to anthropocentrism."*

As to the probability objection against evolution by chance, Monod argues that since the biosphere is a unique event, there is no basis for propositions regarding its likelihood. Actual experiments testing its likelihood would have to involve the creation of billions of evolutionary planets or universes identical to ours in every respect, including the details of history; only then could one determine the probabilities of the events that created human beings. Using Popper's own argument: *propositions regarding Grand Design are not testable; that is, they are not science.*

Evaluating the likelihood of individual steps of the process of evolution is not simple, either, because the events were biased by other than random factors. That is, **the process was directed, not by an animist spirit, but because behavior provided orientation**. Behavior is an obvious element for genetic success in human evolution, but it also applies to earlier lines in our ancestry. The first fish that crawled ashore found that they had to wiggle along the ground or jump rather than flap their fins to propel themselves through space. The different needs incurred by this changed behavior directed the pressures of selection and ultimately led to the evolution of amphibians, reptiles, birds, and mammals.

The argument, used by Eccles, that the evolutionary line leading to conscious human beings is so unlikely that it could not have been driven by chance alone can also be attacked on the basis that the evolution of lions, giraffes, elephants, et cetera, is an equally unlikely series of events. If the improbability of the human evolution is a sign of a Grand Design, what then *is the role of the unrelated players in this process?*

## *Against Eccles' Argument from the Creation of Self*

*Preformation* in biology is the doctrine that every germ cell, the cell from which a new organism develops, contains every detail of the grown-up organism in miniature, so that development is nothing but growth of already preformed parts. The contrasting doctrine is *epigenesis,* according to which every advance of the developing organism is the creation of newness.

Eccles' argument against mechanistic materialism involving the creation of Self rests mainly on the spontaneous epigenesis of complex structures from a single cell, involving the formation of countless differentiated cells, and of the multiplicity of the teleonomic structures of a complex organism. All the cells of a human being originally came from the same single initial set of DNA molecules, and yet they developed differently and matured to radically different teleonomic performances. From the single zygote—the cell formed by the union of two gametes—cell differentiation leads to disparate entities such as muscle cells, brain cells, receptor cells, skin cells, the human eye, and so on. To Eccles, this is a mysterious process that transcends the realm of physics and chemistry, though it is compatible with it.

For Monod the process is little understood, but mysterious it is not. Cell differentiation involves *selective gene activation.* The reactions involved are complex and largely unexplored, but otherwise are normal molecular processes. Among them the spontaneous folding of globular proteins and their agglomerations are molecular models for the spontaneous emergence of macroscopic structures, and the interactions on which they are based very likely are also the source of the autonomous epigenetic differentiation of an organism.

It is now known that globular proteins will fold a random coil in such a way that they will adopt, out of an infinity of possibilities, the single, ordered **native state** in which alone they are able to function. The folded state is stabilized by subtle non-bonded interactions between different parts of the molecule, which are much weaker than ordinary (covalent) chemical bonds. By a slight change in the conditions of the medium in which it exists—for example, by changing its acidity—the folded state can be undone, and protein activity is stopped. However, when the original conditions are restored, the protein refolds spontaneously. *Thus, protein folding is a molecular example of the spontaneous emergence of complex structures from disordered states.*

In addition to folding onto itself, a single globular protein, a *monomer*, can associate with others, forming *oligomers* with a structure of higher complexity. The process is characterized by its *high selectivity*. Agglomeration involves only specific partners in a specific way, to yield the single active conformation that is the purpose of the agglomerate. Such structures arise spontaneously from random mixtures of thousands of proteins. Monod has termed this process *epigenetic*, because it is the emergence of differentiated structures of higher order, very similar to epigenetic embryonic development. The resulting structures depend acutely on properties of the monomers. Thus, such formation of complex structures is not, as he put it, *"a creation but a revelation."*

Ribosomes are larger particles involved in the biochemistry of cells. They are formed by the association of molecules involving dozens of monomer proteins and their weight can reach a million units. By a change in the state of the medium in which they exist, they can be made to dissociate, losing the ability to perform their designated function. As with folded monomer proteins, when the original conditions are restored, the spare parts will reassemble spontaneously and form the same native functional state from which they dissociated. When disassembled parts of T4 bacteriophage, a virus that affects bacteria, are brought together in solution, they assemble spontaneously to the fully functioning system.

In the same way, not only the components of cells but, in tissues and organs, the cells themselves associate with discrimination. This is how Monod rejects the thesis of any non-chemical agent in the formation of a self, proposing that microscopic morphogenesis is model and source of macroscopic morphogenesis, that complex structures *can—even must—*assemble autonomously and spontaneously, and that if most details are unexplained it does not mean they are unexplainable:

> . . . the process of spontaneous and autonomous morphogenesis is based on the stereospecific recognition properties of proteins; . . . it is primarily a microscopic process before manifesting itself in macroscopic structures . . . macroscopic structures (tissues, organs, limbs, etc.) . . . present problems on a different scale. Here . . . the most important constructive interactions occur not between molecular components but between cells. It has been established that isolated cells of a given tissue are able to recognize one another discriminatively and to associate . . . Whatever the case may be, . . . the structure of (recognition) patterns such as these would of necessity be determined by the shape-recognition properties of their protein components. (Monod, 1971)

**In summary: cell differentiation, morphogenesis, the general epigenesis of macroscopic structures from the microscopic, all very likely are based on nothing but highly stereospecific (lock and key) non-bonded molecular interactions.**

## *Rejecting the Animist Covenant*

Regarding the origin of all theories of life which attempt to invoke other than chemical and physical—that is, mechanistic—principles, Monod (1971) notes that *"we would like to think ourselves necessary, inevitable, ordained from all eternity. All religions, nearly all philosophies, and even a part of science testify to the unwearying, heroic effort of mankind desperately denying its own contingency."*

It is a part of the vitalist doctrine that the special life forces in nature are manifestations of purpose. If nature has a purpose, life can have a meaning. This establishes a covenant between human beings and nature. In contrast, mechanism regards physical phenomena as manifestations of objective laws without any purpose for the human fate, and in doing so destroys the covenant and makes life pointless.

During much of our evolutionary history, life in a tribe was the basis of survival. Once the non-human part of the biosphere was controlled to the extent that it did not represent any serious dangers anymore, then strife between tribes, mortal warfare, that specifically human achievement, provided the pressures of selection. At that point cohesion of the horde became a vital principle, and the ability for teamwork and discipline was more important than personal initiative and imagination.

For social animals—ants, termites—civilization is genetically coded. The requisite behavior is automatic. For human beings, who are not automata, cohesion of society is not genetically programmed but supported by culture. To establish and maintain cohesion of the group, explanations are needed. In the course of cultural evolution, these were mostly provided in terms of myths and religions—Monod calls them *"ontogenies"*—which told people why things were the way they were and why one had to act the way one was to supposed to act. As education, they were characterized by one immensely important property: **knowledge and values were derived from the same source**.

In this admittedly simplified account of human history, Monod saw the origin of our craving for explanations and purpose in life. Cultural evolution, albeit not genetically coded, eventually found some genetic support by *"innate categories"* resulting from the pressures of selection. Since the explanations of social order during most of the course of evolution were animist, projecting human nature into the rest of the world, our craving includes the need for a universal teleonomic principle. *"Every living being is also a fossil. Within it, all the way down to the microscopic structure of its proteins, it bears the traces if not the stigmata of its ancestry."*

In this world order of myths and religions, the birth of science—the reliance on objectivity as the sole source of true knowledge—was a disaster. By first testing and then falsifying the explanations given by myths and religions of the order of the universe, science destroyed their authority as a source of knowledge and thus, without intending to do so, shattered the foundation of social stability and the unquestioned authority of the accepted systems of values. With the infallibility of the mythical institutions gone, the problem arose of how to find a new system of values and on whose authority to enforce it.

Cultural evolution is a hazardous process. Like any other evolution, it is the story of change. Mutations engender new ideas in conflict with the old which do not die out on the spot but continue their struggle for survival. While the value of a new idea is recognized by some, a contrasting tradition is still adhered to by others. This is the preprogrammed conflict in an evolving culture and the original sin of modern science. In the scientific world true knowledge is derived from the *"confrontation of logic with experience,"* as Monod put it, in a process that strictly separates knowledge from values. As soon as this principle—objectivity— was accepted as a part of the cultural evolution, it destroyed the animist covenant and its values, demanding *"an agonizing reappraisal of man's concept of himself."*

In this way science pushed man into the *"icy solitude"* of a universe that did not care about human concerns and had no plan in mind for our future. Life now appeared seemingly without meaning, adrift in space-time and going nowhere. This is the root of the *sickness in culture, das Unbehagen in der Kultur,* of societies torn by the inexorable mechanism of cultural evolution: On the one hand, the successful performance of objectivity as the basis of knowledge and technology, and science as the basis of our power and survival; on the other hand, adherence to the old animist myths—in Monod's terms *"soothing, easing the soul, allaying anxiety"* —to preserve coherence in society and to justify what is right and what is wrong, but now without any basis.

This is the story of the broken covenant between man and nature, the evolution of life without meaning, the reason science has never been admitted to the hearts of the masses: it is *"the abyss of darkness"* that has opened before us. In Monod's words (1971):

> The fear (of modern science) is the fear of sacrilege: of outrage to values; and it is wholly justified. It is perfectly true that science attacks values. . . . it subverts every one of the mythical or philosophical ontogenies upon which the animist tradition . . . has based morality: values, duties, rights, prohibitions.
>
> If he accepts the message (of science) in its full significance, man must at last wake out of his millenary dream and discover his total solitude, his fundamental isolation. He must realize

that, like a gypsy, he lives on the boundary of an alien world; a world that is deaf to his music, and as indifferent to his hopes as it is to his suffering or his crimes.

Monod came close to formulating a theory of ideas similar to Popper's and Eccles' concept of world-3. Considering their evolution, he noted that ideas *"retain some of the properties of organisms. Like these, they tend to perpetuate their structures and to multiply them; they too can fuse, recombine, segregate their content; in short, they too can evolve, and in this evolution selection certainly plays an important role."* But, whereas these perpetuated evolving structures and the evolution of Popper's world-3 are similar, the focus is again different—here a warning, there a promise. For Popper and Eccles, world-3 has the immensely beneficial function of offering interactions by which we become selves. For Monod the persisting structures are a sign of danger, representing the darkness of the animist past that threatens the kingdom that science promises.

Is there a Grand Design, a Universal Mind at work in the evolution of the universe and of humankind? After all the surprising discoveries, after all the advances in knowledge and technology, the question is still here—*"not resolved but transformed,"* as Monod (1971) put it: *"It might be thought that the discovery of the universal mechanisms basic to the essential properties of living beings would have helped solve the problem of life's origins. As it turns out, these discoveries, by almost entirely transforming the question . . . have shown it to be even more difficult than it formerly appeared."*

# Part 4

## Divine Reality

# Chapter 5

# THE IMPORTANCE OF THE SELF-CONSCIOUS MIND

## *The Interaction of Mind and Matter: Information as a Causal Agent*

Among the historic responses to the challenge of reality, the frequent reference to the importance of mind is an impressive and constantly recurring theme. At countless times in our history it was the mind that seemed to provide the desired answers for fundamental problems: From Plato's eternal ideas, **St. Augustine**'s "cogito sum," to Kant's claim that the laws of nature are made by the mind, Popper's world-3, Eccles' dualist interactionist hypothesis, and the observer-created reality of quantum mechanics, *experience of the mind, of Self, appeared as the basis of certainty of human knowledge.* The discovery of the principle in modern times by Descartes is a touching document: *"I comprehend, by the faculty of judgment alone which is in the mind, what I believed I saw with my eyes."* And: *"For, in fine, whether awake or asleep, we ought never to allow ourselves to be persuaded on the truth of anything unless on the evidence of our reason. And it must be noted that I say of our reason, and not of our imagination or of our senses."* (*Meditations*, part II, and *Discourse of Method*, part IV.)

From here evolved the Cartesian view that true knowledge resides in what the mind perceives *clearly and distinctly* and the postulate that reality consists of two types of substances: *mind (thinking substance)* and *matter (extended substance),* which reside in two different and disconnected parts of reality. In human beings both substances are combined, but separate. It followed that human bodies are machines, a concept seemingly confirmed when Harvey (in 1628) discovered the circulation of the blood.

From Cartesian dualism arose the problem of how the disconnected worlds interact. How does volition cause motion and stimulus reach consciousness? Questions of this kind have traditionally been an enigma because mind/matter interactions are of a kind that, until recently, differed from everything else that we know. In the ordinary processes of classical physics, it takes some physical intrusion to change the state of a system. Only recently were phenomena of other kinds discovered—*the information-driven observable changes of quantum states.* For reasons of clarity we recall some of the processes, described in part II of this book, in which what was known of a system either touched on the discussion or was directly involved in determining the outcome of an experiment:

- In **photon interference experiments** interference patterns can be destroyed by obtaining information on the paths taken by photons without in any way perturbing them.
- In **quantum-coherence experiments** involving Bell's inequality, information on one particle—obtained by a measurement—can affect the state of a second particle a long distance away.
- In **electron diffraction experiments**, single electrons seem to *know* the state of the entire apparatus and adjust their behavior accordingly. Also, when it is *known* which slit an electron goes through, the interference pattern observed is different from that when nothing is known.
- The **Pauli principle**—*no two electrons in the same system may be in the same quantum state* (see Appendix 13)—is responsible for much of the order in the universe. It implies that, when two molecules approach one another, somehow the electrons in one molecule *know* the quantum states of the electrons in the other. Making use of the term *knowing* in describing such phenomena is, in a way, a manner of speaking, but deliberately so, as Margenau (1984) has pointed out.

Information-driven phenomena convey the impression that the background of reality has **mind-like qualities**. This view is also suggested by the fact that the order of the universe is determined by *probability fields and symmetry principles* which are closer by nature to elements of the mind than of the material world. They are mental concepts, mathematical, not necessarily bound to objects of mass-energy.

In view of phenomena of this kind, A. S. Eddington (1930, 1939) was led to conclude that

> The universe is of the nature of a thought or sensation in a universal Mind. . . . To put the conclusion crudely—the stuff of the world is mind-stuff. As is often the way with crude statements, I shall have to explain that by "mind" I do not here exactly mean mind and by "stuff" I do not at all mean stuff. Still this is about as near as we can get to the idea in a simple phrase. The mind-stuff of the world is, of course, something more general than our individual conscious minds; but we may think of its nature as not altogether foreign to feelings in our consciousness. . . . Having granted this, the mental activity of the part of the world constituting ourselves occasions no surprise; it is known to us by direct self-knowledge, and we do not explain it away as something other than we know it to be— or rather, it knows itself to be . . .
>
> The mind-stuff is not spread in space and time. But we must presume that in some other way or aspect it can be differentiated into parts. Only here and there does it rise to the level of consciousness, but from such islands proceeds all knowledge. The latter includes our knowledge of the physical world . . .
>
> It is difficult for the matter-of-fact physicist to accept the view that the substratum of everything is of mental character. But no one can deny that mind is the first and most direct thing in our experience, and all else is remote inference—inference either intuitive or deliberate . . .
>
> Consciousness is not sharply defined, but fades into subconsciousness; and beyond that we must postulate something indefinite but yet continuous with our mental nature. This I take to be the world-stuff.

## The Importance of the Self-Conscious Mind as a Basis for Knowledge

Whether my mind exists separately from my brain or not, I do not know. Strangely enough, I have a feeling of all kinds of parts of my body—my feet, my hands, my stomach—but no feeling of my brain. Whether my mind is the offspring of blind chance or teleonomic guidance, I do not know. However, it is quite clear that I do have a mind and, above all, I **am** it, and it is of exceeding personal importance that this mind exists before a background that is itself mind-like. The web of probability fields extending through all of space, the flow of non-local, thought-like influences from unknown sources, molecules in superpositions of states: *That the background of reality is mind-like, in all likelihood means that it communicates with other minds.*

## REMOVING THE ILLEGITIMATE BASIS OF SCIENCE

It is the function of empirical science to provide rationally sound and empirically tested knowledge. It is the paradox and predicament of empirical science that the procedures in obtaining such knowledge are not based exclusively on empirical data and rational analyses, but need the involvement of **principles of inference** which are non-rational and non-empirical in the sense that they cannot be verified by an observation of physical reality nor derived by a process of reasoning. Since they have to do with knowledge, in part I of this book we called them *epistemic principles* (from the Greek word for knowledge, *episteme*).

Hume was right: *Causality* cannot be established as a principle of nature by observing external phenomena such as the collision of masses. *Connection* between any two consecutive external events cannot be inferred from their *conjunction*. However, when the self-conscious mind itself is directly involved in a causal link—for example, *when its associated body takes part in a collision, or when the mind by its own free will is the cause of some action—then there is no doubt that causal connections exist.* Thus it is not true that there is *no* experience of causal connection. While it must be admitted that this principle is non-empirical, in the sense that no experience of the external world can establish its validity, experience of causality by the Self is not only that some phenomena can be causally connected, but that they **must be** connected so.

The self-conscious mind also provides the **certainty of identity** or **continued existence**. It is because we ourselves do not constantly vanish from the universe that we assume the same for the cat that gets out of sight behind the sofa. *Not only do we not* *constantly dissolve from reality; we are not able to.* For the same reason, other complex objects cannot do it either, whether demonstrably so by uninterrupted observation or not.

In the same way, viewed from outside, it may not be possible to prove the existence of an **objective outer reality**. However, whenever that body which is associated with this (in some sense) independent self-conscious mind takes part in a process of that outer reality, **experience by the mind of this interaction** leaves no doubt that an objective outer reality exists.

Thus, the first step toward the comprehension of an epistemic principle is an experience of the self-conscious mind. The second step—the one that turns experience into principle—is the specific way human minds react to internal experience. The specific way is part of a *system program* that enables the mind to recognize in its internal experiences the expressions of general principles. The system program instructs us: trust that the future resembles the past, trust that conjunction can mean connection, trust that objects in and out of sight are the same, trust that the experiences of Self are a guide to the true nature of external reality. As shown in Appendices 1 to 4, the program further implements elements of appreciation and principles of aesthetics, such as the preference for simplicity.

Like the program of a computer that does not evolve from its hardware the mind is not the author of its system program, but for all we can see, it has external origins. Thus we have to ask, *where does the program come from?* An answer frequently given is that it comes from evolution, which has programmed us to think in a certain way, because all those perished that did not. But what an amazing view this answer holds, implying that the complex program of mind is coded by the genes, that the thinking of each mind at one point is contained in a few molecules. Apart from the puzzle that this view represents, even if evolution is the answer, then where does evolution come from?

One cannot avoid the feeling that something else is involved, that our ways of thinking are not determined only by the molecules that form our genes. Since the epistemic principles transcend experience of physical reality—cannot be verified by it—and since they transcend human reasoning—cannot be established by it— it is at least possible that they reflect the order of a part of physical reality that transcends the mechanistic foreground of things. Where else to look for the roots of this order than in the transcendent —the mind-like background of physical reality? *The argument set forth here proposes that, as an extension of the mind-like part of physical reality—partaking of a higher order and logic—mind provides the basis of the certainty of knowledge and the foundation of science.*

In this way, identity, permanence, factuality, reality, and causality —all those precious requisites for a reasonable life and for understanding reality, albeit uncertain in experience and reason, can be taken as valid, because they are principles given to us by the independent self-conscious mind. They are not ad hoc hypotheses to make science possible. Rather, they suggest that mind is the extension of a part of reality that transcends the world of mass-energy. These principles are *principles of nature* even though *not anchored in the space-time mechanistic part of nature.* They are, as we have called them, **non-rational and non-empirical, but not mindless**.

In this process, regardless of whether or not the mind exists outside of the brain, the impression of its *independence* is inevitable considering that *operations are performed and principles applied on a level that transcends the level of ordinary material things.* Even though it is based on the brain, new qualities emerge in the self-conscious mind that are not part of the world of the brain and not predictable from its structure, as epistemic principles are not predictable from the mechanics of the material world, and as my feelings and the content of my thoughts are not predictable from the DNA in my cells. There is a level of the brain and another of mind. A new quality arises in the latter that in vain one would search for in the former. In that sense, even if not separate from the brain, the self-conscious mind is the *independent* self-conscious mind.

*Thus, the metaphysical foundations of science are the principles of the independent self-conscious mind.* It is this mind that makes scientific knowledge possible, not by inventing it, but by being part of a level of reality that, by its principles, is higher in hierarchy than the visible part of the world and alone affords true knowledge. In this sense *ultimate reality resides in ideas,* as Plato said. In this sense, *the laws of nature are made by the mind,* as Kant said—(not fabricated by mind, but derived from its connection to a higher reality). This is why Bertrand Russell said, *skepticism is logically impeccable, but psychologically impossible.* **Logic** merely provides the rules to coordinate correctly the processes of reasoning in the brain, making use of the wiring as one should; it refers to something outside of the mind—mindlessly if we are not careful. Any reasoning that is logically correct must at the same time conform to the principles of mind—that is, the order of the universe—to have meaning and to be true. With this in mind, we can return to the outset of this book and redefine the three-fold basis of knowledge:

1. *Human knowledge extends as far as our experience of physical reality* —that is, it must be in agreement with our senses.

2. *Human knowledge extends as far as our reason*—that is, it must be in agreement with the mechanics of the brain.

3. *Human knowledge extends as far as the epistemic principles of the self-conscious mind*—that is, it must be in agreement with the order of the universe.

## The Opening of the Universe

Since the performance of the mind is at a level that transcends the mass-energy physical reality of the brain, it represents an open end to the mechanistic universe and demonstrates the insufficiency of the mechanistic clock as representative of all reality. There is a flow of information, not energy, across a frontier between the independent self-conscious mind and the brain, leading to the concept of an *open universe*. Mind is the hidden variable in the doctrine of mechanism and its actions entail that the world of mass-energy is not completely sealed.

An open-ended universe is also suggested by the quantum phenomena. As when changes in information can cause changes in physical phenomena; as when non-local and non-material influences can have instantaneous effects on the quantum states of distant systems; as when virtual particles can appear out of nothing; as when an electron in diffraction experiments acts as though it were at the same time everywhere in the extended apparatus, transcending the world of localized mass points, opening a window, as it were, to a different state of reality. *Because of the uncontrollable effects of non-local influences, the present is open. Because of the role of Heisenberg events in creating reality, the future is open.*

Thus, there is growing evidence *by elementary physical phenomena* for an openendedness of the universe that Newton's machine once closed. The world of mechanism is just the cortex of a much wider and deeper reality. The classical mask has cracked and the classical physical reality has sprung leaks from which emanations of a different way of being are seeping in, out of a background that has mental properties. All the following statements have become acceptable: *the basis of the material world is non-material; the components of ordinary things are not mass-energy real in the same way as the things that they make; the constituents of ordinary things that have shape do not themselves have a definite shape; the permanent order of the visible world is based on transitory and chaotic processes; in the dual-slit experiments with electrons we have found access to the world of Platonic ideas.*

The assumption of the self-conscious mind as an extension of the mind-like background of the universe may be able to shed new light on a number of unsolved puzzles, such as *the amazing finality of the evolutionary process; the amazing formation of a Self from a single cell; the phenomenon of the unity of perception; the phenomena involving Jung's archetypes and collective subconscious; the temporal and topical parallelism in world-3 developments; the conception of empirical concepts prior to the discovery of the corresponding phenomena.* The human mind often anticipated empirical principles prior to their discovery—Plato's atoms as forms, Aristotle's potentia, Democritus' atoms as building blocks of matter, Berkeley's "esse," Zeno's paradoxes, are some examples. **The overwhelmingly important faculty of the human mind: it can be inspired by unknown sources** as though it were in contact with a part of reality that transcends the level of mass-energy.

What can the opening of the universe suggest? If the universe is non-local, we must expect that human beings necessarily are a part of the web of effective influences that are the basis of its nature. If the universe is made of *"mind stuff,"* we must expect that human beings are connected with it and that the organ by which the connection is made is the human mind. If the nature of the universe is mental, we must assume that it includes us and interacts with our minds. Why is the experience of mind so important? Because it is the experience of the kind of stuff the universe is made of. If something mental is at work in the universe, it is a natural consequence that the Self is its extension.

In addition to interacting with our mind, there are other channels by which the background can access our most intimate properties. The nature of living beings is molecular; information for the physiology of a human being is laid down in a number of molecules. The duplication of the genes, at the moment of conception, is a quantum process subject to quantum uncertainty and non-local superposition effects. How the background is involved with the genetic transcription, or that it is involved at all, is not known. At the same time, what is amazing is not that the background of the universe is affecting us, but the extent of the local control that we are allowed to exert.

*The level of reality, beyond the foreground of its mechanistic order, is determined by factors to which we must ascribe omnipresent, infinite, boundless, instantaneous, universally pervasive, and almighty efficacy—attributes usually associated with Divine Reality.*

Thus, after a long period of mechanistic darkness, our scientific views allow again for a spiritual basis of the nature of humanity. It is not true that science is the basis of atheism, and a false accusation that the two are synonymous.

The Search for Divine Reality is, of course, nothing new. In fact, it has been the concern of religions through the ages. What *is* new is the fact that the search can now be performed within the framework of the physical sciences, and not in constant conflict with them. Visions that Newton's mechanics once blinded have become possible again.

These, then, are the essential functions of the self-conscious mind as they appeared so far: In our experience of reality, it provides the unity of perception by the synthesis of fragmented stimuli. In our development as a species and Self, it is a sensor to an open universe, to a different state of reality, to a network of non-local effects. In our reasoning, it provides the unity of understanding by the epistemic principles without which a reasonable life and understanding of the world would be impossible.

We can now turn to arguments which support the view that, in addition to the epistemic principles, the mind also provides the ethical principles whose acceptance alone can put us in harmony with the rest of the universe. Thus, the discovery of a coherent universe that is essentially akin to human nature—of a mind-like character—reveals itself as a most wonderful gift, establishing again a basis for a life with meaning and the potential to transcend the limited world of physical existence.

# Chapter 6

# HEALING
# THE WOUND

## Objective Knowledge in a Life with Values

Socrates related virtue to knowledge. While he did not mean *knowledge about physical reality*, I am willing to take the additional step: ***what we know about physical reality must affect our way of life.*** A certain morality is connected with our knowledge of reality, a certain conduct is compatible with its nature and with our understanding of it. Responding to the challenge of physical reality can guide the mind, as experience is translated into a basis on which one can act. This is the connection between epistemology, ontology, and ethics: *if we know and understand, we can choose to be good.*

I will summarize again Monod's view (1971) of the effects of science on the traditional values of society: Traditional societies based their social order on myths, *"animist ontogenies,"* religious systems, that justified human values in terms of explanations of the nature of things and the origin of the world. By assuming a divine purpose in the universe, they gave meaning to life. By giving an account of why things are the way they are, they were able to dictate rules of conduct that were in accordance with the body of knowledge that they purported to afford. *Thus, knowledge and values derived from a single source, and the rules that they established formed a consistent set.* When the infallibility of the animist ontogenies was shattered by scientific objectivity, the very foundation of social order, of the meaning of life, and a life with values seemed dissolved. By attacking the ontology of the animist theories—their astronomy, biology, physics, and chemistry—science inevitably also attacked their values.

Thus arose the *"lie at the root of the sickness of spirit"* of modern societies as they cling desperately to values and myths whose foundations have been destroyed by the principle of objectivity at the base of technology. Monod's conclusion is that a radical reappraisal of the traditional values and the rejection of many of them is unavoidable.

*Such a reappraisal can now be undertaken, for a life with values and meaning is no longer in conflict with the principles which serve us as the true source of knowledge. It is not the rejection of traditional values that is needed now, but the reconstruction of a foundation on which they can be based.*

## *The Significance of the Epistemological Paradox*

The epistemological paradox (Monod, 1971) is the contrast between the admission of teleonomy in the biosphere and the postulate—at the basis of science—that nature is objective. Adherence to objectivity precludes any reference to final causes and purpose in nature. Monod's paradox: **Scientific objectivity forces us to conclude that living organisms incorporate purpose, versus: objective science must exclude purpose from its descriptions of nature.** Monod's presentation of the matter is particularly instructive:

> There is no physical paradox in invariant reproduction . . . : invariance is bought at its exact thermodynamic price, thanks to the perfection of the teleonomic apparatus . . . This apparatus is entirely logical, wonderfully rational, and perfectly adapted to its purpose: to preserve and reproduce the structural norm. And it achieves this, not by transgressing physical laws, but by exploiting them . . . It is the very existence of this purpose, at once both pursued and fulfilled by the teleonomic apparatus, that is the miracle. Miracle? No, . . . there is really no paradox or miracle, but a flagrant epistemological contradiction.
>
> The cornerstone of the scientific method is the postulate that nature is objective. In other words the systematic denial that "true" knowledge can be reached by interpreting phenomena in terms of final causes—that is to say, of "purpose." . . . the postulate of objectivity is consubstantial with science, and has guided its prodigious development for three centuries. It is

impossible to escape it, even provisionally or in a limited area, without departing from the domain of science itself.

Objectivity nevertheless obliges us to recognize the teleonomic character of living organisms, to admit that in their structure and performance they decide on and pursue a purpose. Here . . . lies a profound epistemological contradiction. In fact the central problem of biology lies with this very contradiction, which, if it is only apparent, must be resolved, or else proved to be radically insoluble, if that should turn out indeed to be the case.

Monod's paradox is central to the biosphere and a most profound revelation. Before any other issues are entertained, we have to explore its significance.

Specifically, we have to consider what exactly the meaning is of this contradiction: *Objectivity by definition must systematically exclude finality in nature, but has to admit it for living organisms.* Does it mean that living organisms are special, requiring more than the objective analysis normally practiced by physics and chemistry? Does it establish, in spite of the many protestations, a hidden vitalism? Or, if living organisms are not special in being teleonomic, should we extrapolate to things outside the biosphere? That is, if some structures of the universe are admitted to serve a purpose, might not others also? *If scientific description of the biosphere is granted inclusion of purpose, why must it be forbidden to other parts of physical reality?*

It would be tempting to conclude that living beings *are* special, since there is no sense of teleonomy in physics nor any need for it, but no living organisms exist without it. However, the suggestion of finality also often arises in cosmology; it is usually met with considerable suspicion, but *perhaps it is possible that in the same way in which the human mind is the manifestation of the mind-like principles of physical reality, the teleonomic structures of living beings are manifestations of the purposefulness of a much grander scheme.* Both have not sprung from nothing: consciousness not from a universe devoid of mind and teleonomy not from a universe devoid of purpose.

Monod (1971) seems to have thought that the paradox would eventually be resolved by the full exploration of the actions of simple chemical principles, such as the stereospecific non-bonded interactions between proteins, which underly all spontaneous morphogenesis. But no matter how thoroughly explored, I do not believe the paradox can be resolved. It will remain because the appearance of purpose in an objective world, no matter how it intruded upon it, is something alien, a dissonant chord, an element from a different order than pure objectivity can embrace. In this sense, while not special, *we are different—we are more than mechanism would allow.* Thus, one might argue that the purposefulness of living beings is in fact a manifestation of kinship with a universe whose principles act at a higher level than those of the simple foreground of things.

The epistemological paradox is echoed in others that pervade our lives. Among them the *paradox of knowledge:* elements of knowledge are rational and empirical—while non-rational and non-empirical principles are needed for deriving elements of knowledge. The *paradox of personal freedom:* the best protection of personal freedom lies in the willingness of individuals to curtail their civil rights. Following the same pattern, the basis of material things is non-material; living organisms are made up of dead atoms; intelligent beings are made up of stupid molecules; our attempts to establish a system of perfect justice have led to a system of technicalities which are corrupting justice; the best rule of government is democratic—the deciding of expert matters by a vote of the uninformed.

In the discussion of these matters, we see that same principle in action again and again; that is, essential aspects of central human concern—regarding the nature of reality, knowledge, morality, purpose in life—all have to admit of an element of uncertainty, of faith, of an unverifiable remnant which suggests that the human mind is not self-contained, that perhaps it is open to a transcendent order, and responsive to a superior Mind.

# The Self-Conscious Mind as the Basis of Ethics
## ON MORAL STANDARDS AS PRINCIPLES OF THE MIND

Now that the authority of the animist myths has been found lacking, the most pressing problem, described well by Monod in *Chance and Necessity,* is this: *on whose authority are the ethical principles to be based that society needs to maintain its order?* The answer is: **on the**

**authority of mind.** *In the same way that the experience of Self, the self-conscious mind, endows the epistemic principles with certainty, it endows the ethical principles with authority.* From what do the epistemic principles derive their certainty? Neither from the experience of external phenomena, nor from a process of reasoning, but from the system program of the self-conscious mind. *By being an extension of the mind-like background of physical reality, mind gives them certainty.* In the same way, from what do the ethical principles derive their authority? Neither from the experience of external phenomena, nor from a process of reasoning, but from the system program of Self, the self-conscious human mind. *By being an extension of the mind-like background of physical reality, mind gives them authority.* Mind can serve this function and establish these principles because it partakes of an order that transcends the mechanistic surface of physical reality. Due to the very nature of these principles and their origin, they form a link between the level of human order and the transcendent order of the universe. Epistemic principles give us a sense of *what is true and what is false.* Ethical principles, of *what is right and what is wrong.* Reflecting the order of the universe, the former establish the certainty of *identity, permanence, factuality, causality;* the latter, of *honesty, morality, responsibility,* and *purpose* in a mind-like reality.

Like their kin, the moral principles are not derived by a process of logic nor verified by an experiment, and *it takes a mind to establish their certainty.* The first step in establishing the authority of ethical principles—as with the certainty of epistemic principles—is an internal experience of the self-conscious mind. For example, it is the satisfaction felt in expressing a truth, or the quiet enjoyment of doing right by others. The second step is the specific way in which the mind recognizes general principles in particular experiences. Thus, the ethical principles are associates of the epistemic principles and, like their kin, they are related to a transcendent part of *physical* reality. This means that violating any one of them, no matter whether epistemic or ethical, will put us in contrast to the nature of reality. **If the background of the universe is mind-like, it can be assumed that reality has a spiritual as well as a physical order, and it is in human minds (to paraphrase Eddington) that this order rises to the level of morality.** *To live in accordance with the essence of things,* as Socrates said, *is the premise of the moral life. One cannot live in peace of mind without at the same time being in harmony with reality.*

The very term "peace of mind" means to live in accordance with the principles of mind, or with the most intimate order of physical reality. It is the sensation of personal identity, or personal involvement in a causal act, that gives the epistemic principles certainty. In the same way it is the *sensation of personal responsibility* that gives the moral principles authority. As there is no other way to *verify* the former, though they are undeniably valid, the latter are valid even though not verified in any way.

Thus, all our metaphysical convictions—the epistemic principles as expressions of the physical order of the universe, and the ethical principles as expressions of its spiritual order—are found rooted in the same mind-like background of the universe, whose existence is being revealed to us by the quantum phenomena. As an extension of this transcendent part of physical reality, and partaking of its higher order, the mind provides the basis of the certainty of knowledge, the foundation of science, the safe grounds on which to base objective rules of conduct, and the only chance for a life in harmony with the nature of physical reality.

The exact meaning of that term, *spiritual order of the universe,* can benefit of some additional explaining. There are two sets of principles—epistemic and ethical—whose validity we trust, but cannot verify, and whose authority we accept, but cannot prove. Since the principles of knowledge supply us with the laws of physics, everyone will readily accept that they are somehow rooted in the physical order of the universe. In our recent history we have not suspected that the principles of ethics, too, could somehow be rooted in the order of physical reality, because the universe has always been that machine, that mindless mechanism in which eternal matter is dominated by eternal laws. Now, however, that the mind-like nature of the universe has been revealed, the story is different, and one cannot help but think that the universe has an order that, at the human level, relates to the spiritual as well as the physical. This is what I take to be its spiritual order.

The concept is really not so amazing as at first it might sound. Assuming that our metaphysical convictions are rooted in the mind-like background of the universe is the same as assuming that they have a chance to be true. To be meaningful, our metaphysics must make a connection with some structure of the universe. This I take to be its spiritual structure.

The validity of the epistemic principles is justified by their success. The question is then, whether a similar element of success can be found for the ethical principles that will support the argument that they, too, are signs of a transcendent order? The answer is that *their success lies in the ability to achieve peace of mind.* Epistemic principles reflect the determinism of physical reality, the ethical principles allow for a choice. Violating the laws of physics and chemistry will make our bodies sick. Violating the laws of ethics will make us restless, homeless, seeking violence and criminal behavior.

## On the Foundations of Ethics in the Mind-like Order of the Universe

It is in their common roots—the mind-like background of physical reality—that the close ties between epistemology, ontology, and ethics come to the fore. Since the *sensation of knowing* is the state of mind that is most familiar to us, it can teach valuable lessons regarding conditions—like those of morality—that are less well defined. For example, when the unexpectedly complex nature of the elements of knowledge is recognized, that is, their dependence on seemingly illegitimate principles and elements of taste and tradition, the ensuing rejection of naive realism can be a catharsis in everybody's life—a liberation—because one realizes that the step can be taken without forsaking objective knowledge. In the same way it is a healing experience when the rejection of naive animism is found to be possible *without forsaking the premises of a moral life based on a covenant with reality.*

The first step in establishing a system of ethics consists of choosing a cardinal value on which everything else can be built. In the light of the data presented in this book, I propose that the resolution *to live in accordance with the order of the universe* is such a fundamental value. Once accepted, everything else will fall into place because *all* the principles of the mind, epistemic and ethical, are accepted as expressions of universal order and guidelines for a moral life.

The system of ethics based on universal order defines a consistent, objective, and comprehensive set of standards of conduct; *consistent,* because the supporting principles are derived from the same source as the epistemic principles; *objective,* because the supporting principles are expressions of the transcendent order of the universe; and *comprehensive,* because it contains propositions such as Monod's *ethics of knowledge* in addition to observing the spiritual order of the universe. Monod's ethics of knowledge is the system of rules of conduct whose foundation is *"the assent to the principle of objectivity as the sole source of knowledge"* (Monod, 1971), and the commitment to *authenticity* as the only basis for a decent life. It is important to emphasize that assent to objectivity and authenticity is a necessary part not only of science but also of the moral life, because no other principles can form a credible basis for human relations.

In Monod's definition "no discourse or action is to be considered meaningful, *authentic,* unless it makes explicit and preserves the distinction between the two categories (objective knowledge and subjective values) that it combines." In the definition of authenticity proposed here, no discourse is authentic unless it is in agreement with all the aspects of the three-fold basis of knowledge—that is, the *data,* the *laws of logic,* and the *epistemic principles;* no action is authentic that is in violation of one of the epistemic or ethical principles of the mind; and *no life is authentic* that is in conflict with the order of the universe. The striving for authenticity—that is, adhering to the principles of the mind as guidelines of one's life—forms a covenant with the mind-like background of physical reality and enables a life in harmony with nature.

*The ideal of authenticity is a wonderful guide of human life.* Striving for authenticity enables a life that is genuine, not counterfeit; it creates a person who is trustworthy, not devious. Connections to the ideals of accuracy, sincerity, honesty, truthfulness, and related values are readily made. It is a sign of the sickness of current societies that public discourse and actions all too often are inauthentic, not genuine but corrupt.

If standards of conduct are not in agreement with the order of nature, they cannot be expected to be effective. In any society moral principles can be effective only if everybody can reasonably agree to them. Nobody can be expected to adhere to principles that are in conflict with our knowledge. Thus, striving for a life in harmony with the order of the universe demands assuming personal responsibility for being reasonably well informed, *enlightened*, and for making an effort to overcome the complacent views of common sense. *Do not adopt as a fundamental standard of conduct any view of the world that is in clear contrast with what is reasonably known about the nature of the universe.*

Considering the fact that all knowledge contains elements beyond its control, it is logical to exercise *caution* in our conclusions. In the realm of human relations, *modesty and tolerance* are the equivalent. Considering that certainty of knowledge— perfection—is impossible, the *impossibility of perfection* must also be accepted in other human concerns, such as in personal relations, democracy, personal rights, and ethics. In the realm of conduct, the *willingness to forgive* is the equivalent. And it ought to be said here that, considering the importance of our kinship with the mind-like background of the universe, a prime goal of *teaching* must be to make young minds receptive to the subtle messages that might be directed at them.

The rules of the physical order of nature are inexorably enforced, but the rules of the spiritual order of nature do not assert themselves in the same way. Their actions are subtle. They are there for human beings to accept *of their own free will.*

## THE MAXIM OF HARMONY

As we cannot act in contrast to the physical order of the universe, we cannot live in conflict with its spiritual order. The good life is in harmony with the nature of the universe.

The maxim of harmony advises us to be inquisitive, not ruminative. Uncertainty in knowledge—*is what I think true?*—allows for creativity. Uncertainty in ethics—*is what I do right?*—allows for free will. *There is no virtue in conflict with what the mind can know about physical reality.*

The rejection of final causes by classical physics was the rejection of their justification by myths. The thesis—*all being is explained by the power of its end; the present is explained by its future*—is so utterly untestable that it was the first one to get into conflict with the newly discovered principle of objectivity.

At this point, the principle of purpose deserves another look. Its rejection cannot be based only on the fact that it is unverifiable. The epistemic principles share the same characteristic, yet they form the foundation of science. If it is true, as set forth earlier, that experiences of the Self can be the source of convictions regarding validity, then the sense of purpose experienced by many human beings may suggest that it is a genuine principle of nature. Since every aspect of the apparatus of a living being is *"the materialization of purpose,"* it may be impossible to excuse the mind from that same principle. Quite apart from any verification, it appears that those who adopt a purpose in life live a happier and more fulfilled existence than those who do not.

Thus it is proposed that *accepting a purpose in life is in harmony with nature*. Some teleonomic structures are dictated to us—like the organs of our body. Others can be adopted for greater enrichment by our own free will.

Monod thought that assent to the ethics of knowledge was incompatible with the craving for purpose, but the opposite is possible. Even though we are part of a physical order that is ruled by causal laws, we can have a free will. Even though we are part of an objective reality, we can adopt a purpose in life.

## The Transformation of Hume's Fundamental Problem of Ethics

At this point Hume's fundamental problem of ethics is transformed in the same way in which his problem of epistemology was transformed above.

Hume's fundamental ethical problem is the thesis that, what *ought* and *ought not* to be done, cannot be deduced from what *is* and *is not the case*. That is, propositions connected by *ought* and *ought not* cannot be deduced from those connected by *is* and *is not*. The former express a relation or affirmation that the latter do not contain, like *necessary connection* is not contained in *temporal conjunction*.

In contrast to Hume's thesis it is proposed that the human mind translates the spiritual order of the universe into moral laws in the same way in which it translates the physical order into physical laws. The order of the universe is what is the case. The moral laws deduced from it tell us what ought to be done.

## ℭe Restoration of the Covenant

We have said that the epistemic and ethical principles are expressions of a higher order than that of the limited human horizon, that this is the order of the transcendent background of physical reality. A system of ethics based on the authority and verity of these principles *forms a covenant with nature*.

The animist covenant was the alliance of kinship. By projecting essential aspects of the human psyche into inanimate objects, investing a soul into rivers, mountains, trees, and all of nature, human beings found themselves in a world to which they were related, a cognate reality that was a protective home in spite of the dangers that it harbored. The covenant had a reassuring effect and any existence outside of it was inconceivable, the senseless life of the homeless. When science turned to objective analysis of empirical phenomena, destroying the basis of knowledge of the animist world, it broke the covenant and opened a painful wound. In Newton's mechanistic universe there was no room for God— *La Place to Napoleon: Sir, I have no need for this hypothesis*—and the misunderstanding of mechanism as a model for all realms of reality turned life into a meaningless exercise.

Monod believed that the need for the covenant is inborn, a genetically supported craving for meaning in life, because we are the descendants of animists, and *"have inherited the need for an explanation, the profound disquiet which forces us to search for the meaning of existence. That same disquiet has created all myths, all religions, all philosophies and science itself."*

In contrast to Monod's hypothesis, I propose that the need for the covenant, the craving for explanations and purpose in life, is not the result of an evolutionary process, but the need of the mind to reach out to what is its kin in the universe, to the mind-like background of reality.

At this point the immense significance of the discovery of the mental qualities of the universe becomes apparent. Twentieth-century science offers the possibility of restoring the covenant while at the same time retaining objectivity as the sole basis of knowledge.

Recognition of the mind-like part of the universe, of **the presence of Mind** (Polkinghorne, 1991), is *the assumption of a universal teleonomic principle* and an animistic view. The covenant between the human mind and the mind-like background of reality is similar to the historic animist covenant except that its assumption is suggested by testable descriptions of elementary phenomena, such as those involving the diffraction of particles, the quantum probability fields, photon-coherence experiments, and Bell's non-locality. Historic animism was also inspired by the experience of external reality, the streaming of rivers, the thundering of mountains, or the growing of trees, but the conclusions drawn—of the projections of elements of the human mind into the inanimate world—were not testable. Still, even with this difference, *the historic covenant is yet another example of an anticipated concept, not correct when its archaic form is taken verbatim, but containing a truth nevertheless.*

Though not animist in the traditional sense, **the modern covenant has the same healing power**. It provides a home again to the homeless, restores meaning to the meaningless life. Its comfort lies in the confidence that we have not been equipped with a mind for no reason. This is the *"millenary dream"* come true: delightful messages are waiting for those who choose to listen, eliminating the *ennui* of a life without purpose.

The need for the covenant is not the result of our descendence from animists, but the need of the mind to live in touch and in harmony with the nature of reality. The sickness in culture is not the sickness of the animists having to live in a world of objective science, but the sickness of those who have cut the covenant with the transcendent part of physical reality. There is a longing for its music, and the disquiet when the messages stop. There must come upon us now a feeling of triumph and liberation in the discovery that we can live with the assent to objectivity and authenticity and at the same time enjoy the blessing of a covenant. There is great joy and peace of mind in the awareness that we are not useless accidents, but realizations of a superior essence.

This is how the two extremes are seen to merge—the reductionist view represented by Monod, and the idealist view represented by Eccles. On the one hand there is the system of ethics that is committed to objectivity, rationality, and authenticity, as bases of what Monod calls a *normative epistemology.* On the other hand, there is the expanded view of reality, as prompted by the quantum phenomena, which instructs us to include in our analysis of the human condition the mind-like background, the epistemic and ethical principles, the higher level of order beyond the mechanistic foreground of things. One does not have to be an animist in the traditional sense to acknowledge the mind-like character of reality. One does not have to be a vitalist in the traditional sense to believe that there is transcendency and more to living organisms than the current laws of physics and chemistry. One does not have to be a separatist to believe that the mind is independent of the brain and that it is the organ that communicates with the mind-stuff of the universe.

Thus, the objectivity of science and Monod's *"confrontation of logic with experience"* as the driving forces are not *"cold and austere,"* do not cause *"aggravating anxiety."* While they may end the old covenant between man and nature, they are establishing a new one. There is no *"abyss,"* but *knowledge and values have found a common source again*—the mind-like background of physical reality and the principles of its physical and spiritual order—and the conclusion is *not* that *"man at last knows that he is alone in the unfeeling immensity of the universe, out of which he emerged only by chance"* (Monod, 1971). On the contrary, there is a consistent system of ethics and knowledge without solitude because our music is the music of the universe, and Mozart is, as Küng wrote, *a touch of metaphysics.* The realistic understanding of the nature of knowledge and the experience of the quantum world suggest that the nature of the mind is not the product of a *"giant lottery"* but, rather, the realization of *universal potentia,* a direct manifestation of the essential character of the universe, and an offer to share.

Sickness of spirit is life in disharmony with the order of nature. It is the sickness induced by the lack of commitment to simple authenticity. In societies that are sick, governments abandon authenticity for the sake of political correctness, public agencies abandon authenticity for the sake of bureaucracy. In societies in which the commitment to authenticity has been abandoned, it is found natural that commercial advertisers misrepresent the merits of their products, that professional athletes actively seek out chances at "good"—that is, unnoticed—breaches of rules, that administrators of educational institutions profess educational excellence by words but not substance, and that entire industries thrive on corrupting society by glorifying violence, dishonesty, hatred, and debased behavior.

In this drama, it will be important that those responsible for the advance of religion avoid the closed mind of the mechanistic sciences. It is the spirituality of science that there is revelation in every new discovery. Science devoid of that spirituality is dead; religion devoid of knowledge is not authentic. In a society in which science and religion are in conflict, something has gone out of balance and needs a tune-up. It is hard to believe that a religion can be truly pious and follows God's will when it is in contrast to the laws revealed in the order of God's universe.

We have spiritual needs because the nature of the universe is spiritual and not because we are some aberration of evolution or the product of an illegitimate phylogeny. Since we are a part of this world we feel the need to be a *functional part* in close connection with it and in agreement with its nature. This is the wonderful promise of being able to be at the same time logically consistent and inspired by a transcendent world. We can combine all the good instincts and anticipations of the classical era—the desire for rationality, accuracy, objectivity, compatibility with experience —with the opening of our minds to Plato's world, a world of purpose, meaning, and truthfulness in a wider sense.

At the end of our search for *Divine Reality*, we have not found an old man with a white beard, sipping mead, and being fluttered about by a flock of beautiful angels, but we find that questions can be meaningfully asked and answers attempted within the framework of science and not in opposition to it, which traditionally has been the task exclusively of religion. *Does the human mind reside outside of the brain*, as Eccles thought? Is there a Grand Design, the *"presence of Mind,"* as Polkinghorne put it? *"Is there,"* as Sherrington asked, *"evidence derivable from Nature that implies the existence of God?"* The answer is that the signs of the transcendent nature of the universe are overabundant. At this point the ancient covenant between man and nature is obsolete—*now is the new covenant between human minds and the mind-like background of the universe.* In the quantum phenomena, mind turns into matter—*the word is becoming flesh.* Insofar as our nature is molecular, we are subject to the non-local quantum effects of the universe. Insofar as it is mental, we are subject to the activities going on in the mind-like background of the universe.

*Two things fill my mind with ever-increasing admiration and reverence the more I think about them: the miracle of my consciousness and its covenant with the mind-like background of physical reality.*

# Epilogue and Summary

# ON THE FOUNDATIONS OF METAPHYSICS IN THE MIND-LIKE BACKGROUND OF PHYSICAL REALITY

That the basis of the material world is non-material is a transcription of the fact that the properties of things are determined by quantum waves—probability amplitudes which carry numerical relations, but are devoid of mass and energy. As a consequence of the wave-like aspects of reality, atoms do not have any shape—a solid outline in space—but the things which they form do, and the constituents of matter, the elementary particles, are not in the same sense real as the real things that they constitute. Rather, left to themselves they exist in a world of possibilities, *"between the idea of a thing and a real thing,"* as Heisenberg wrote, in superpositions of quantum states, in which a definite place in space, for example, is not an intrinsic attribute. That is, when such a particle is not observed it is, in particular, nowhere.

In the quantum phenomena, we have discovered that reality is different from what we thought it was. Visible order and permanence are based on chaos and transitory entities. Mental principles—numerical relations, mathematical forms, principles of symmetry—are the foundations of order in the universe, whose mind-like properties are further established by the fact that *changes in information* can act, without any direct physical intervention, as causal agents in observable changes in quantum states. Prior to the discovery of these phenomena, information-driven reactions were a prerogative of the mind. *"The universe is of the nature of a thought. The stuff of the world,"* Eddington wrote, *"is mind-stuff."*

Mind-stuff, in a part of reality behind the mechanistic foreground of the world of *space-time energy sensibility,* as Sherrington called it, is not restricted to Einstein locality. The existence of non-local physical effects—faster-than-light phenomena—has now been well established by quantum coherence-type experiments like those related to Bell's inequality. If the universe is non-local, something that happens at this moment in its depths may have an instantaneous effect a long distance away; for example, right here and right now. By every molecule in our body we are tuned to the mind-stuff of the universe.

In this way the quantum phenomena have forced the opening of a universe that Newton's mechanism once closed. Unintended by its creator, Newton's mechanics defined a machine, without any life or room for human values, the Parmenidian One, forever unchanging and predictable, *"eternal matter ruled by eternal laws,"* as Sheldrake wrote. In contrast, the quantum phenomena have revealed that the world of mechanism is just the cortex of a deeper and wider—a transcendent—reality. The future of the universe is open, because it is unpredictable. Its present is open, because it is subject to non-local influences that are beyond our control. Cracks have formed in the solidity of the material world from which emanations of a different type of reality seep in. In the diffraction experiments of material particles, a window has opened to the world of Platonic ideas.

That the universe should be mind-like and not communicate with the human mind—the one organ to which it is akin—is not likely. In fact, one of the most fascinating faculties of the human mind is its ability to be inspired by unknown sources—as though it were sensitive to signals of a mysterious origin. It is at this point that the pieces of the puzzle fall into place. Ever since the discovery of Hume's paradox—*the principles that we use to establish scientific knowledge cannot establish themselves*—science has had an illegitimate basis. Hume was right: in every external event we observe conjunction, but infer connection. *Thus,* causality is not a principle of nature but a habit of the human mind. At the same time, Hume was wrong in postulating that there is *no* single experience of causality. When the self-conscious mind itself is directly involved in a causal link—for example, when its associated body takes part in a collision, or when the mind by its own free will is the cause of some action—then there is a direct experience of, and no doubt that, causal connections exist. When this modification of the paradox is coupled with the quantum base, a large number of pressing problems find delightful solutions.

Like the nature of reality, the nature of knowledge is counter-intuitive, and not at all like the automatic confidence that we have in sensations of this phenomenon. The basis of knowledge is threefold. The premises are *experience of reality, employment of reason, and reliance on certain non-rational and non-empirical principles*—the assumption of permanence and identity, factuality, causality, and induction. Where do these principles come from? Neither from an experience of external phenomena nor from a process of reasoning, but from a system program of the self-conscious mind. *By being an extension of the mind-like background of nature and partaking of its order, mind gives the epistemic principles—those used in deriving knowledge—their certainty.* Since they are not anchored in the world of space-time and mass-energy but are valid nevertheless, they seem to derive from a higher order and transcendent part of physical reality. *They are, it can be assumed, messages from the mind-like order of reality.*

**It is the same way with moral principles.** Traditional societies based their social order on myths and religious explanations. By assuming purpose in the world, they told people why things are the way they are, and why they should act the way they were supposed to act. In the *"animist ontogenies,"* values and knowledge derived from a single source, and life had meaning in an *"animist covenant,"* as Monod called it. By destroying the ontological base of the animist explanations—their astronomy, physics, and chemistry—science also destroyed the foundations of their values. In this process Monod saw the origin of the contemporary dilemma: on the one hand, science is the basis for our power and survival; on the other, it has broken the animist covenant, rendered life meaningless in the process, and disconnected the world of values from the world of facts.

This sickness of spirit and the concomitant erosion of moral standards are the great dangers for the future of humanity, already apparent in the public adoration of violence and debased behavior. At its roots is the unsolved question: on whose authority are the moral principles to be based now that the authority of the animist myths has been found lacking? **For those who are willing to listen, the answer is: on the authority of mind.**

*In the same way that the self-conscious mind grants certainty to the epistemic principles, it invests authority in the moral principles.* Like the former, the moral principles are non-empirical and non-rational—not derived by a process of logic nor verified by experience—messages from a higher reality beyond the front of mass-energy sensibility. Epistemic principles give us a sense of what is *true* and *false;* moral principles, of what is *right* and *wrong.* The former establish the certainty of *identity, permanence, factuality, causality;* the latter, of *responsibility, morality, honesty.* By the same process that allows us to accept, without possible verification, the epistemic principles, we can also accept the authority of the moral principles. Violation of any one of them will put us in contrast to the nature of reality. If the nature of the universe is mind-like, it must be assumed to have a spiritual order as well as a physical order. As the epistemic principles are expressions of the physical order, the ethical principles are expressions of the spiritual order of physical reality. *By being an extension of the transcendent part of nature and partaking of its order, mind establishes the authority of the ethical principles.*

The challenge of reality and the ability to explore it are wonderful gifts to humanity. Understanding reality requires refinement of thought. That is, it has to do with **culture**. It requires an effort, is not afforded by automatic, intuitive reflex. Making sense of the world takes the response to a challenge, not the complacency of common sense. It is the same as striving for the moral life. An important part of it is the need to become aware of the specific character of human nature, to recognize *"the human mystery,"* as Eccles called it; the mystery of how mind and body interact, how self-conscious human beings with values emerged in an evolutionary process supposedly based on blind chance and brutality. The evidence is growing that there is more to human nature than the laws of physics or chemistry, that evolution has not by mere chance brought us to a place where we live, as Monod wrote, "at the boundary of an alien world that is deaf to our music and indifferent to our hopes and suffering and crimes."

The barbaric view of reality is mechanistic. It is the easy view of classical science and of common sense. In *epistemology,* mechanism is *naive realism,* the view that all knowledge is based on unquestionable facts, on apodictically verifiable truths. In *physics,* mechanism is the view that the universe is clockwork, closed, and entirely predictable on the basis of unchanging laws. In *biology*, mechanism is the view that all aspects of life, its evolution, our feelings and values, are ultimately explicable in terms of molecular properties. In our *legal system*, mechanism is the view that the assumption of precise procedural technicalities constitutes justice. In our *political system*, mechanism is the view that the assertion of finely formulated personal rights constitutes the ideal democracy. In our *public administration*, it is the view that responsible service manifests itself by the enforcement of finely split bureaucratic regulations. All of these attitudes are the attitudes of barbarians.

The quantum phenomena have taught us that, without naive realism, knowledge is possible. They have taught us that, without naive animism, assent to objectivity as the sole source of knowledge *and* a life with values are possible. Principles exist which are valid even though they cannot be verified. The discovery of the quantum phenomena has established a new covenant—*between the human mind and the mind-like background of the universe*—one that provides a home again to the homeless and meaning to the meaningless life. Whether or not the human mind is separate of the brain, as Sherrington and Eccles thought, I do not know. But I do not doubt that it is human only in some parts, and in others shares in that mind-like background of the universe. It is now possible to believe that the mind is the realization of universal potentia, a manifestation of the essence of the universe, whatever name we give to that essence. Therefore, *the good life is in harmony with the nature of physical reality.*

# THE ILLEGITIMATE COMPONENTS
# OF KNOWLEDGE

## *The Importance of Aesthetics and the Insufficiency of Observation in Deriving Knowledge*

The historic dispute involving the Ptolemaic and Copernican systems that scandalized Europe during the outgoing Middle Ages and early Renaissance clearly demonstrates that observation is important for our scientific convictions, and, at the same time, unimportant.

The issue concerned the center of the universe. In the Ptolemaic system, the earth is the center; in the Copernican system, the sun. In the late Middle Ages in Europe, the Ptolemaic system was adopted as part of Church dogma, and objective testing and critical analysis of the issue at that time carried serious risks.

In the third century B.C. **Apollonios of Perge** and in the second century A.D. **Ptolemy of Alexandria** explained the movements of the solar system in terms of the *eccentric, epicycles,* and *deferents.*

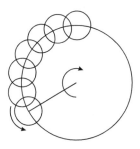

FIGURE 5
*The Ptolemaic system*

Every planet moves in an *epicycle,* the small circle in this figure. While it revolves in this orbit, the center of the epicycle moves in a larger circle (the *deferent*) whose center (the *eccentric*) is close to the position of the earth. Positions of the epicycle at different times of a year are shown. The actual motion of a planet is the composite motion—the superposition of the two cyclic motions.

In this model of the solar system, called Ptolemaic not after its inventor, the earth is the center. Planets move in *epicycles (epi, meaning upon, besides, on the outside of)*. Centers of epicycles move in other circles, the *deferents* (those that carry on). The double system moves around a center, *the eccentric*, which is close to, but does not coincide with the earth. The actual motion of a planet is the composite motion—the superposition of the two cyclic motions.

Using this computational scheme, tables of planetary constellations were compiled that were in agreement with the data of the time. With accumulating observations, discrepancies appeared, but they could be removed by placing a second epicycle on the first, a third on the second, and so on, up to sixty, seventy, and more, for a single planetary orbit.

Due to the lack of sufficiently sophisticated computational instruments, the Ptolemaic system soon became hopelessly complicated and inconvenient to use. In addition, it became non-unique; that is, systems with different epicycles could be constructed for a single planetary orbit. At this point the Ptolemaic system lost all its credibility as a model of reality. Interestingly, it had started out as a computational device. Only later, when it worked with success, was it considered a model of reality.

In the system proposed by **Copernicus** (1473–1543) in the sixteenth century, the Sun is the center of the universe, and the planets revolve around it in circles. The Copernican system really originated with **Aristarchus of Samos** in the third century B.C. At that time, however, the concept was rejected because it did not fit the generally accepted religious views. Science must always conform to the style of a time and, when ideas are premature, true or false is not the issue.

The circular nature of celestial motion originally was an essential feature, because only circular motion was considered perfect—leading nowhere—and nothing but the most perfect quality was accepted for the objects of the sky. As it turned out, this aspect of the Copernican system was in error. Some fifty years after Copernicus' death, **Johannes Kepler** (1571–1630) found that circular planetary orbits are not in agreement with the data. Rather, *the planets are moving about the Sun in ellipses.*

It is a popular myth that, in the sixteenth century, the Ptolemaic system was rejected because the Copernican system agreed better with the data. This view is not correct. Both models agreed equally well with the existing observations; preference for one over the other was based on different considerations. Some, nevertheless, were ready to kill, others to die, to assert their view of "reality."

The killings and torture were all the more tragic **because the whole question was meaningless.** First of all, not the center of the universe was at stake but, at best, the center of our solar system. There are millions of suns in the galaxy, millions of galaxies in the universe; neither the Sun nor the earth can meaningfully be claimed to be the center of all.

Second, from a *position in space at rest with respect to the sun*, the planets will be seen to revolve around it. From a *position at rest with respect to the earth*, the planets and the Sun are seen to revolve around the earth. All motion is relative to a point of reference that is a matter of definition.

Interestingly enough, the **motions observed,** which look like superpositions of epicyclic orbits, are called the **apparent motions.** The motions not observed—the ellipses—are called the real motions. It is true that the Ptolemaic system is not in accordance with our general understanding of physics today, while the Copernican system (or Kepler's modification of it) is. But our current knowledge of physics was not available at the time of Copernicus, and the evidence did not then exist for refuting one theory and establishing the other. No, the real reason why Ptolemy was rejected and Copernicus accepted lay not in the accuracy of the planetary predictions but in value judgments of a deeper kind—*it was solely because the latter was the simpler theory of the two.*

*Simplicity is often invoked in scientific disputes, on the principle that, of two competing scientific theories, the simpler one is the better one.*

Simplicity is of course not a scientific principle but a matter of taste and aesthetics. Nevertheless, the propensity for simplicity is important in scientific issues, because somehow the mind has been programmed to look for simple connections rather than complicated ones. *When observation is incomplete (which it usually is) with the missing aspects supplied by the mind, out of an infinite number of possibilities the simplest one is assumed.*

In the same way, the belief in the *identity* of things and their *permanence* is a matter of simplicity. If everything were to vanish from the universe when out of sight, to reappear only when observed again, things would be terribly complicated.

No principle of logic, no single observation vitiates the opposite view, that nature is complex and, hence, *complicated theories are better than simple ones.* That the cat that crawls under the sofa at one end is not the same cat that, after some time, reappears on the other side, is not less logical an assumption than the contention that it is. In fact, one would be hard put to prove either.

A lawful nature is simpler than a chaotic one. Faith in lawfulness and the immutability of physical laws is faith in simplicity. It is a value judgment, one of many automatic biases induced by the system aesthetics of the mind.

In agreement with this aesthetic program, simplicity is considered a virtue in many applications. For example, **William of Ockham** (1285–1350) formulated the principle that we know as **Ockham's razor,** a symbol of simplicity synonymous with *rational clarity* and *economy of thought,* which is accepted now as the most satisfying mode of thinking and expressing one's thoughts. The razor is found in Ockham's writings in various formulations, such as *"To employ a number of principles when it is possible to use a few, is a waste of time."* Or, *"we must never assume a number of elements unless we are forced to do so."* William of Ockham is also of interest here because he proposed to *separate logical reasoning from theological reasoning* when the Church attacked his teachings on dogmatic grounds. Thus he destroyed the union between philosophy and theology, between science and religion, facts and values, and helped to create the world of **double standards** that we live in today.

The **psychology of vision** provides additional indications of the disposition of mind for simplicity. As noted by **Arnheim** (1964, p. 179), when two dimensions of a surface are seen in a graph, and the observer is asked to complete the figure by adding a third dimension, *out of a large number of possibilities automatically the simplest one is chosen.* A circular surface seen from the top, for example, is conceived to be smooth, like a taut drumhead, not like a slack drum or one with bruises in it.

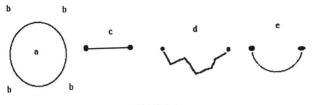

FIGURE 6

In the figure above, *a* tends to be seen as a surface that lies on top of plane *b*. In theory this surface could assume any one of an infinite number of shapes, just as the straightness of a drumhead is only one of the innumerable shapes that we could obtain by draping a tablecloth over the drum. Yet, the surface in the circle is perceived as straight as in *c*, and not as curved in the manner of *d* or *e*.

This example is of interest also because it shows that visual perception can be ambiguous, in spite of the feeling of certainty that we may have when we look at something. Other well-known examples of deceptive perceptions include perceptions of the size of rectangles under different angles of projection, or the display of a homogeneously gray surface against a background of varying brightness.

Bias in perception also derives from the fact that the perceptive machinery has to some extent been adapted to the needs of a species. For example, Lorenz (1966) quotes a discovery by **H. B. J. Barlow**, who found that a relay station between a frog's eye and its brain allows the transmission of signals of a fly in motion, but not of a fly at rest. The image of a resting fly on the retina is simply not perceived, and only victims in flight will be considered as prey, to avoid the rotten meal.

## Rational Beauty

In a fascinating analysis Polanyi (1958) has pointed out how, among the aesthetic factors that guide the promotion of a theory to the status of established truth, the appreciation of **rational beauty** is an important one. Historic details involving Einstein's Special Relativity make an excellent case to document the power of this principle.

In 1887 Michelson and Morley measured the speed of light in direction of the motion of the earth around the Sun and perpendicularly to it. It is usually reported that the same speed was found in each direction. Thus, the conclusion followed that *the speed of light in free space is the same everywhere, regardless of the motion of the light source or observer. As it turns out, this account is inaccurate.*

The constancy of the speed of light inspired Einstein to formulate his theory of special relativity and, as shown in Appendix 4, it is a fundamental postulate on which this theory is based. The story goes that, in 1905, when Einstein learned of the Michelson-Morley experiment, it was this encounter that prompted him to formulate his famous theory. *As it turns out, this account is also inaccurate.*

The inaccurate version of the history of relativity theory is usually reported because it reflects the conventional view of science. According to this view, science is supposed to advance in this way: in the first step of a scientific development, experiments are performed that yield some unexpected results. In the second step, applying the principle of induction, the observed results are used to derive a new theory. Thus, experiments inspire theory; verifications of theory establish truth. This is the naive view of physical science.

In contrast to the conventional perception, science does not proceed in this way. Specifically, when Einstein conceived the theory of relativity, he did not even know of the Michelson-Morley experiment. Rather, the cause of conception was a *rational event.*

As he describes it in his biography, Einstein discovered relativity *"after ten years' reflection . . . from a paradox upon which I had already hit at the age of sixteen: If I pursue a beam of light with velocity c, I should observe such a beam as a spatially oscillatory electromagnetic field at rest. However, there seems to be no such thing, whether on the basis of experience or according to Maxwell's equations. From the very beginning it appeared to me intuitively clear that, judged from the standpoint of such an observer, everything would have to happen according to the same laws as for an observer who, relative to the earth, was at rest."*

As to the alleged confirmation of Einstein's theory by the Michelson-Morley experiment, one has to consider that the speed of the earth's orbital motion around the Sun is approximately 30 km/s. In their experiments, Michelson and Morley expected to find an effect of this magnitude in the speed of two light beams, one propagating perpendicularly to the orbital motion and one, in direction of it. As the story is usually told, no such effect (or drift), **zero,** was found. **Not quite true.** Rather (for details, see **Polanyi,** 1958), a drift **of 9 km/s was found,** and reproduced by **Hicks** and **Miller** in repeated experiments between 1902 and 1926. When Miller's results were reported to a meeting of the American Physical Society in 1925, did everybody rush to abandon Einstein's theory of relativity? No. The theory is much too beautiful, its rational clarity too convincing, to be easily abandoned.

In his book *Personal Knowledge* (1958), Polanyi described the events in the following way:

> The layman, taught to revere scientists for their absolute respect for the observed facts, and for the judiciously detached and purely provisional manner in which they hold scientific theories (always ready to abandon a theory at the sight of any contradictory evidence), might well have thought that, at Miller's announcement of this overwhelming evidence of a "positive effect" in his presidential address to the American Physical Society on December 29th, 1925, his audience would have instantly abandoned the theory of relativity. Or, at the very least, that scientists—wont to look down from the pinnacle of their intellectual humility upon the rest of dogmatic mankind—might suspend judgment in this matter until Miller's results could be accounted for without impairing the theory of relativity. But no: by that time they had so well closed their minds to any suggestion which threatened the new rationality achieved by Einstein's world picture, that it was almost impossible for them to think again in different terms.

As it turns out Miller's observations were later explained without abolishing the theory of relativity, and the constancy of the speed of light in a vacuum eventually was experimentally established, but not by the technique used by Michelson and, in any case, decades after the formulation of relativity theory (Polanyi, 1958).

*The example shows that science (what we can know) is not simply based on experimental results (experience of reality). Observations of particular events (like the fossil record) will not always lead to the abandonment of a general conviction (like creationism). The rationality of a theory and its beauty are also of overriding significance. We know nothing about the world if we do not first observe it. At the same time, we know more about it than can be observed.*

# Modes of Thinking by Tradition

No inventory list of the illegitimate components of knowledge would be complete without an analysis of the fixed patterns of thinking that tradition has deposited in our minds. Among them we find the traditions of **materialism, mechanism,** and the **Platonic** and **Aristotelian traditions**. The awareness of the hidden powers of these modes of thinking over our mind is a prerequisite for any critical analysis of thinking. The automatic life does not allow the full potential of its creative powers. From conscious thinking we learn the conscious life.

### THE TRADITION OF MATERIALISM

By searching for the *original substance,* the *primal element, Urstoff,* out of which all ordinary things are formed, the early Greek philosophers founded a powerful tradition of thinking that has remained effective to this date: the tradition of **materialism,** the belief that everything that exists is of a material nature. As **Margenau** (1983) phrased it: *"Materialism as a doctrine is based on the proposition: To be is to be material; only matter exists."*

The Greek philosophers typically used *reasoning* to arrive at insight into reality (see, e.g., Russell, 1946, 1978). It is a strange aspect of reasoning that repeated thinking of *what is possible* will lead to *convictions of how things are.* For example, *since data are not like the objects, since experience of reality is not that of the true nature of things but only of the appearance of things, we can meaningfully ask if the very existence of objective reality cannot be denied?* If that line of reasoning is repeated over and over again, all in a sudden the question may subtly change into the assertion that *there is no objective reality, only the ego and its sensations exist (solipsism).* There is an equivalent technique used in rhetoric: repeated exclamations of an assertion establish its proof.

The Greek philosophers could have performed experiments to test the results of their reasoning, but they opted not to. Most of them were members of a class of gentlemen and all practical—experimental —work was left to slaves. In the European Middle Ages, this tradition continued. Theoreticians were associated with church schools and communicated in Latin. The experimentalists were the craftsmen and practical inventors who lived in cities and spoke the vernacular. Members of these two social classes found it hard to communicate.

Even today, the tradition persists. Experimentalists in science often find it hard to respect theoreticians, and a sharp line is drawn between practitioners in business and industry and representatives from academia. The contempt of the experimentalists for the church schools lives on in our language. An important part of the curriculum in the schools, the *"artes liberales,"* was the *"trivium,"* the *"threefold path,"* consisting of grammar, rhetoric, and dialectic. Todays meaning of *"trivial"* is the revenge of those who were excluded.

## THE PLATONIC TRADITION

In the philosophy of **Plato** (427–347 B.C.), eternal truth resides in immutable forms, eternal **ideas** (see, e.g., Russell, 1978). Plato's ideas are not just general concepts, but are truly real; they are, in fact, the ultimate reality. They are prototypes of particular things, primal images existing outside of space and time. They form a reality that is transcendental, beyond this world. Each particular concrete thing is a copy of the eternal idea of that thing, but an inferior copy, belonging to a lower state of reality.

The Platonic tradition implies that understanding, knowledge, and truth must be self-consistent, final, unchanging. Since the world is filled with objects of inferior reality, knowledge of real things—the ideas—is primary, and empirical science of secondary importance. The transcendent ideas are not perceived by the senses, but possibly are comprehended by intuition. *The soul remembers.* Understanding is not a matter of experience, but insight. Interesting polarities mapped out by this view are those between *thinking and perceiving, reason and the senses, ideas and visible things, reality and appearance.*

There are similarities between the Platonic tradition and the earlier doctrines of the Pythagoreans. **Pythagoras of Samos** (c. 540 B.C.) proposed that *not stuff, but numbers* are the basis of the world. The harmony of the *cosmos* (a term he invented) rests on divine and eternal order based on the ratios of numbers, like quantum numbers, as harmony in music is based on the ratio of frequencies. Thus, in this philosophy *we find non-material, mental attributes at the basis of reality.*

The experience of mathematics inspired the belief in exact and eternal truth. Geometry deals with ideal circles and triangles. No real circles are exactly circular, no real triangles exactly triangular. It seems logical to conclude that, in general, all exact reasoning applies to ideas, not real things.

The harmony of the cosmos, Pythagoras claimed, could be "heard," *by contemplation, not sensation.* There was a consensus that sense perception somehow is not suited to the search for real truth.

## THE ARISTOTELIAN TRADITION

In contrast to Plato's transcendentalism, Aristotle thought that ideas, or forms, are not transcendental but exist in the things (see, e.g., Russell, 1978), that true reality is **immanent, not transcendent.** In the Aristotelian view, there is no transcendent idea of a horse, existing somewhere, of which a particular horse is a copy. There are no transcendent forms of things, but the forms are present in the things. Animals, trees, have a soul *"psyche."* Nature is animated, the universe an organism. (And now in quantum mechanics the universe is characterized by *"wholeness";* it is an organism.)

According to Aristotle, a tree has a soul that guides its growth; soul gives unformed matter its form; the form is the goal and purpose of the evolution of a tree, as all natural phenomena, even physical ones, occur for a purpose. Question: *Why does a stone drop to the floor?* Aristotle: *Because that is where it belongs.*

Taken verbatim, the theories of the Greek philosophers often sound too speculative and inexplicable to be convincing. Nevertheless, they are expressions of basic attitudes which are still dominant in our thinking and founded powerful traditions of thinking which are difficult to escape.

Around A.D. 1000 in Europe, Platonism was the accepted doctrine of the Catholic Church. Romanesque Art, evolving at that time, reflects the Platonic tradition. In the early Romanesque churches, light is a symbol of the presence of God. It enters into the sacred space through a small number of tiny openings which are not visible from anywhere in the building because they are hidden in an upper level—transcendent as it were—behind a triforium. As true reality, the reality of God is outside our reach, transcendent in the Platonic tradition, removed from the pilgrims who see the light but from where they know not.

During the twelfth and thirteenth centuries, the worldview changed. Aristotle's philosophy was rediscovered and finally became part of church dogma. Nearly simultaneous with this development, Gothic architecture evolved with walls of buildings resolved into windows, often stretching from the clerestory to the floor. Thus, in contrast to the Romanesque churches, the pilgrims are standing *in the light*. Divine reality is **in** this world, immanent in the Aristotelian tradition.

Generally, when eternal truth is not in the things, but transcendent in a different world, when things are imperfect copies of eternal ideas, the visible surface of things is not so important as the hidden essence and realism in art is not the desired style. Thus, expressionism, surrealism, any movement of abstract art (aimed at revealing the essence of things, the immutable ideas below the visible surface) are expressions of the Platonic tradition. In contrast, when the essence is in the thing, when the form of an object is the expression of its soul, then the visual appearance is important. Thus, any realistic movement in art is an expression of Aristotelian tradition.

It is often said that Romanesque sculptures are non-realistic because the sculpted figures were subjugated to the demands of architecture. The case is perhaps more symbolic than technical: art was an expression of the general views of the time and, as the human form imperfectly reflected reality, the non-realistic representations of people, often imprecisely termed "grotesque," were attempts at revealing their hidden essence. Gothic sculpture, in contrast, is typically life-like, a transformation of Aristotelian views. (See Hofstädter [1988] for additional details.)

## THE TRADITION OF MECHANISM

The mechanistic worldview is a transcription of Newton's mechanics. It is a classic example of how traditions persist in our thinking and, coming from distant sources, often vigorously reemerge in entirely different environments. The mechanistic doctrine rests on three postulates (Bohm, 1957): (1) *All phenomena can be explained by the mechanics of moving particles;* (2) *Moving particles obey Newton's laws;* (3) *Causality is a principle of nature.*

The trajectory of a particle in classical mechanics consists of its position and momentum as functions of time. If these variables are known exactly at a given moment, then from a knowledge of the forces between the particles, the infinite future and infinite past of the system can be calculated. Thus, if everything is determined by the mechanics of moving particles, and if their motions are exactly predictable for the infinite future, the future of the whole universe can be predicted, provided all the relevant data are known at one time. This is the ideology of **Determinism,** the thesis that everything is set, to the extent even that human beings have no free will. In a mechanistic universe, when the data are known, nothing unexpected can happen. *The classical universe was a machine, exactly predictable, clockwork, a closed system.* If reality is only what mechanism admits, God has no place in it.

A clockwork universe means that everything has its own precisely defined attributes, acts as a predictable cause, like a gear in the mechanism of a clock. The future is closed, because everything is predicted. The past is closed because it can be precisely constructed from the present. The present is closed because there are no agents unaccounted for (uncaused), no instantaneous, unexpected, non-local influences. **Heisenberg's uncertainty principle** (see part II) has destroyed the basis of this doctrine. According to this principle the position of a particle and its momentum cannot both be known exactly at the same time. In fact, all variable physical properties come in pairs; when one is known exactly, the other is unknowable. Thus, on principle, all the data cannot be known that a mechanist needs to predict the future of the universe.

In many ways, the mechanistic worldview is the expression of common sense. For a long time, it was widely considered the "modern view," and Newton the founder of "modern science." In actuality, the doctrine is a new formulation of ancient ideas and traditions of thinking.

To **Parmenides** (c. 450 B.C.), *"to be"* meant *"to fill space solid, to be material."* In addition, he considered true reality as *immutable,* not allowing for any change. Again, if the Parmenidian ideology seems arcane and less than convincing, Newton's mechanics, absolutely familiar and acceptable to us, is Parmenidian materialism in disguise. (See the excellent summaries by Heisenberg [1962], Sheldrake [1988], Davies and Gribbin, [1992]). In Newton's mechanics, material particles are the **real** things, *"solid, massy, impenetrable."* And in accordance with Parmenides' concept that *there can be no change, there is no becoming,* in Newton's universe **"eternal matter moves in accordance with eternal laws,"** as Sheldrake put it (1988). Matter, energy, and the laws of nature are eternal. In the physical eternity of the mechanistic universe, there is no becoming, no real change. The infinite future is contained in the past.

The industrial revolution, inspired by Newton's mechanics, was materialism in practice. In the social order of the times, clockwork humans were cogwheels in a giant machine, with inert minds that responded in a predictable way to external stimuli (as in the mechanistic context of behaviorism). In our times, a new kind of economy seems concomitant with a new kind of physics. As quantum phenomena reveal the immaterial basis of the material world, industry is shifting from the production of goods to the transfer of information.

Isaac Newton (1642–1727), in his *Opticks:*

> All these things being consider'd, it seems probable to me, that God in the Beginning form'd Matter in solid, massy, hard, impenetrable, moveable Particles, of such Sizes and Figures, and with such other Propensities, and in Proportion to Space, as most conduced to the End for which he form'd them: and that these primitive Particles being Solids, are incomparably harder than any porous Bodies compounded of them: even so very hard, as never to wear or break in pieces; no ordinary Power being able to divide what God himself made one in the first Creation . . . And therefore that Nature may be lasting, the Changes of Corporeal Things are to be placed only in the various Separations and new Associations and Motions of the Permanent Particles.

The desire for the unchanging has religious roots (Sheldrake, 1988). In epistemology, the same desire becomes the craving for certainty of knowledge. In a changing world, the Greek Atomists were in search of the unchanging. The Pythagoreans were in search of God. The roots of our attempts to understand the nature of reality are spiritual. **Davies** and **Gribbin** (1992) trace the concept of unchanging truth to the experience of the unchanging self. "I" is the experience of **"all is one"** and **"nothing changes."** If everything is in flux, there can be no "I." Since we are aware of "I," we conclude that nothing meaningful changes. We grow older and our bodies are altered, but the "self" to which this aging body belongs stays the same.

*The laws of nature are independent of the things that they regulate.* They are almighty, unchanging, ubiquitous, and nothing is hidden to them or outside their power. Sheldrake (1988): *The ancient idea that truth resides in immutable forms, ideas, continues with the immutable physical laws of the universe.* Like Plato's ideas, the physical laws are independent of the matter that they regulate. Outside of this world, they are the **eternal truth, the ultimate reality.** *Did they exist before there was a universe?* Heinz Pagels asked (1982).

Toward the end of the middle ages in Europe, Aristotle's teaching displaced Plato's in church dogma. In the Renaissance, the Platonic/ Pythagorean tradition was rediscovered. Sheldrake (1988) quotes **Nicholas of Cusa** (1401–1464), who considered the world as infinite harmony: *"Number was the first model of things in the creator's mind."* And Copernicus: *"Things that are mathematically true are exactly true."* And Galileo: *"Nature acts in accordance with unchanging laws, which it never violates. . . . The book of the universe is written in mathematical language. Its symbols are triangles, circles, other geometrical shapes, without which we could not understand a single word and would helplessly err around in a dark labyrinth."*

The concept of "traditions of thinking" is very similar to the concept of "frame of thought" in Eddington's philosophy. In *Philosophy of Science* (Eddington, 1939) various frames of thought are characterized. A special frame of thought, for example, is the definition of knowledge as a description of reality. Another is the concept of analysis, the conviction that a whole can be understood by taking it apart.

*The material presented above shows that our thinking is affected by traditions of culture. Everyone's thinking follows some tradition. None of your thinking originated with you. Attitudes deposited by tradition are yet another automatic element of thought, another illegitimate ingredient in the process of thinking, usually taken for granted and applied uncritically.*

Traditions of thinking are often reflected by long-lived ideals of education. For example, for centuries the ancient Greeks adhered to the ideal of *Kalokagathia* (from *kalos* = beautiful; *kai* = and; *agathos* = noble, or good), the idea that the moral education and development of the soul advance the beauty of a person's body; that to be good is to be beautiful. In Roman times the same ideal became *mens sana in corpore sano*, a sound mind in a sound body. Similarly, the traditions of *oida me eidos* and *scio me nescire*, meaning "I know that I know nothing"—or, perhaps, "Knowing nothing, I still know something. Not knowing I know."

## THE PROGRAM OF THE MIND

In dealing with elements of knowledge we are *not* objective, but guided by elements of appreciation, aesthetics, and fixed patterns of thinking that culture has deposited in our mind. Instinctive inclinations and emotional values are in place before the mind is entered by signals from the realm of facts. All this is apart from the fact that our thinking is ruled by the chemistry of the mind and the logic of the wiring.

To repeat the most important point: the human mind is a complex program of automatic operations that are not under its self-conscious control. Principles accepted without doubt (the permanence of things, causality, the laws of common sense), traditional attitudes automatically adhered to (like the doctrines of mechanism or materialism) hidden propensities (for simplicity, rational clarity, lawfulness of nature), and the dependence on senses which censor and impregnate the very signals they send to the mind, are part of an operating system whose origins are not now obvious to us. Again and again we are faced with the fact that the human mind is not a self-contained device, *but an open end of our universe.*

The hidden principles deposited in our mind are of three different kinds: *epistemic principles*—those that have to do with deriving knowledge and understanding reality; *aesthetic principles*—those that define automatic propensities applied in making decisions; and *ethical principles*—those that define the rules of conduct.

If the issue is to think clearly and distinctly (that is, realistically), then it is important to identify what, in our thinking, is under our control, and what is part of the automatic machinery.

# THE FAILURE OF INTUITION

It is commonly said that our thinking is in phase with reality, because biological evolution has forced it to be so. **Ernst Mach** (1838–1916), the Austrian physicist and science philosopher, was a prominent proponent of this view. Mach believed that we are in command of instinctive knowledge, common sense, which typically tells us what cannot occur. It is a sense of what is absurd; it tells us what will not happen, rather than what will happen. Since biological evolution has adapted our mind to reality, our mental patterns are natural patterns.

The problem with this view is that reality is counterintuitive. Common sense is in accord with some of reality, but not with all of it. Einstein's relativity and the phenomena of the quantum world are fair examples of elements of reality contrary to common sense. In order to illustrate this point, some aspects of Einstein's special relativity theory will be shortly reviewed.

## *The Conceptual Background of Relativity*

All motion is relative. Thus, statements about motion are meaningful only with regard to some frame of reference. For example, a person in a moving railroad car can at the same time be called *at rest* (namely, with respect to the railroad car) and *in motion* (namely, relative to an observer standing on a station platform). Einstein's special relativity has to do with whether or not an **absolute frame of reference** exists in the universe.

The question arose in various contexts, among them the problem of the motion of light in a vacuum. At the end of the nineteenth century, light was generally considered to consist of *electromagnetic waves*, periodic changes of an electric field coupled with changes of a magnetic field. When electric charges oscillate, they generate such changing fields, which can then detach themselves from the oscillating charges and propagate through space as electromagnetic waves. Since all known waves exist in some medium—sound waves in air, for instance, or water waves in water—while light can travel in empty space, scientists at the end of the nineteenth century believed that all of space was filled with a light-bearing medium called the **ether.** Its existence seemed necessary for the propagation of light, but the problem was that no one had ever found any evidence for its existence.

In 1887 **Michelson** and **Morley** performed an experiment aimed at detecting the luminiferous (light-bearing) ether. The experiment showed, to the surprise of many, that the ether does not exist, that light can propagate in a vacuum, and that its speed in empty space is a constant, independent of how fast the light source or observer may be moving relative to one another.

The Michelson-Morley experiment was based on the fact that, when a river flows at a certain speed, it takes more time to send a boat downstream and up, than the same distance across and back. If the universe is filled with ether, which is stagnant in absolute space with respect to some absolute frame of reference, then the orbital motion of the earth around the sun (~30 km/s) must create a rapid ether stream. By comparing the time needed for a lightwave to travel a certain distance across this stream and back with the time that it needs to travel the same distance downstream and up, existence of the ether, and the rate at which we move through it, should be demonstrated.

Michelson and Morley measured the time it took for lightwaves to travel a certain distance in the direction of the orbital motion of the earth and back, and the same distance perpendicular to that orbit and back. When the measurements were actually made, surprisingly they *indicated no difference.* Since no ether stream was detectable in this way and no other measurable properties of this medium can be found, Michelson and Morley concluded that the ether does not exist. They also postulated that *the speed of light in a vacuum is constant,* at 186,000 miles per second.

Einstein's special relativity theory is related to the Michelson Morley experiment because it is based on these two postulates:

(1) The speed of light in a vacuum is constant and independent of the motion of light source or observer.

(2) The laws of physics are the same in all frames of reference that move with constant velocity relative to each other.

These two statements may seem rather harmless, but they have corrupted everything that common sense ever suggested concerning the nature of space, time, and matter.

## The Relativity of Time

Consider a railroad car with a light source (**S**) placed exactly in its center. The train is in motion and passes an observer (**O-1**) who is stationary outside on a station platform. A second observer (**O-2**) is seated inside the car. At a certain moment, two flashes of light are emitted from **S** in direction of two detectors mounted at the front and rear ends of the car. One of the flashes is traveling from the center toward the front, and the other toward the back. The purpose of the experiment is to determine what the observers **O-1** and **O-2** will see.

From the point of view of **O-2**, inside the railroad car, **both flashes will strike their targets at the same time.** This is so, since they originate exactly in the center of the car, they have to travel exactly the same distance to the rear- and front-end detectors, and they are pursuing both paths with identical speeds (again, independent of the motion of light source or detector).

In contrast to this, from the point of view of **O-1** on the station platform, it appears that the detector at the rear end of the railroad car is moving *toward* the signal that is approaching it, whereas the front-end detector is *receding* from the lightwave emitted in its direction. Thus, for **O-1**, **first the rear end is hit, and then the front.**

The question now is, who is right? **O-1** or **O-2**? The answer is that both observers are right, each one within his frame of reference. The conclusion is that *two events that are simultaneous to one observer need not be so to another.* Contrary to common intuition, *simultaneity is relative.*

Other counterintuitive conclusions easily follow from the two postulates of relativity. For example, when a spaceship leaves the earth at a speed of 180,000 miles per second, and a light pulse is sent after it, the speed with which that pulse approaches the ship is 6,000 miles per second *as measured by an observer on earth,* but 186,000 miles per second *as measured by a person in the ship.* The first is the difference, speed of light minus speed of the rocket, as determined on earth. The second is reported by the observer in the rocket because the speed of light is a constant, regardless of motion between source and observer.

How can it possibly be that both observers are right? They are right within their respective frames of reference, **because motion affects the properties of measuring instruments.** Specifically, in the direction of motion, **moving yardsticks contract** and, compared with stationary clocks, **moving clocks are slow.**

## The Slowing of Clocks by Motion

A particularly simple clock that illustrates the effect motion can have on clocks consists of two mirrors mounted at the end of a glass rod whose length is precisely known. A light pulse reflects back and forth between the two mirrors, striking each of them at a precise interval— the time of flight along the length of the rod. Each time one of the mirrors is struck, the device records a tick.

When such a clock is in motion relative to an observer, the timing pulse is seen by that observer to travel a longer distance between two ticks, compared to a stationary clock. This is so, because the motion of the pulse between the two mirrors is added onto the motion of the mirrors in space. Since the speed of light is constant, longer distance of travel corresponds to longer time of flight; that is, the ticking of the clock is delayed by its motion. Thus, compared with stationary clocks, moving clocks run slow.

Since the laws of nature are the same in all frames of reference moving at constant speed with respect to each other, *all clocks must be affected in the same way,* including biological ones, even though the cause of delay may not be so obvious. According to this, people in motion age differently from people at rest, since motion affects the rate of metabolism.

## The Relativity of Mass

In classical physics, the mass of an object is an absolute property, immutable, eternal, and indestructible. In reality it is not. As one of the consequences of Einstein's special relativity it appears that mass, m, and energy, E, are equivalent: $E = mc^2$, where c is the speed of light in a vacuum.

The formula implies that basically mass and energy are the same thing. Thus, **whenever the energy of an object is changed, its mass is changed, too.** For example, the higher the speed with which an object moves with respect to some observer, the higher its energy and the bigger its mass relative to that observer.

*Hot water is massier (heavier) than cold. A loaded spring is heavier than that same spring relaxed.* When hot water is cooled, energy is given off. That energy is paid from its mass content. When a loaded spring is relaxed, the energy it spends comes from its mass content.

In our ordinary experience the statements of relativity do not make sense. This is so because, in all but a very restricted part of reality, *common sense is common non-sense.* Classical physics was a systematic transcription of common sense. It is based on a number of fundamental concepts, such as space, time, and matter, which are reliable and independent *categories* in the world of human consciousness. Upon a closer inspection of reality, their absolute nature and independence were lost.

Space and time are connected and depend on point of view. There is no absolute, universal time, but time depends on where one is. Clocks are slowed by motion. Moving objects shrink. Simultaneity is relative. Matter warps space and can be annihilated. A loaded spring is heavier than the same spring relaxed. People living at sea level age more slowly than those dwelling on mountaintops. *All these propositions are violations of that faculty we are so proud of—common sense.*

## The Local Nature of the Universe

Relativity of mass is connected with the concept of the **locality of the universe.** In mathematical terms, the dependence of mass on velocity is expressed as $m = m_0/(1 - v^2/c^2)^{1/2}$. Here $m_0$ is the mass of an object when it is at rest, the so-called *rest mass,* while m is its moving mass—the mass apparent to an observer relative to whom it is in motion with velocity v; and c is the speed of light in a vacuum.

It is seen from the formula that v cannot exceed c, because then the square root would be of a negative number, which is impossible. In addition, when v = c, m becomes infinite. This case is also impossible, because an infinite amount of energy would be needed to move an infinite mass. Thus, *the consequence of relativity is that no material object can attain the speed of light or exceed it, and that the fastest speed possible for any signal is the speed of light.*

This is the origin of the concept of the **local nature** of the universe. Events are local when they cannot be affected by non-local influences which have traveled from somewhere else at more than the speed of light. Thus, in considering physical phenomena on earth, I do not have to be concerned about any processes occurring now in the depths of the universe. Vice versa, in order to have an effect on the sun, for example, I must start my operation at least eight minutes prior to the desired time of action, because that is the minimum time required for any signal to travel from the earth to the sun; to cause some action now on a star that is a few million light years away, my preparations here must have been made a few million years ago.

In contrast, **non-locality** of the universe implies that something I do now, here, can have an immediate effect at any other point in the universe, no matter how remote it is. Relativity seems to prescribe that non-locality is impossible, that the universe is local, at least in the sense that no signals can be transmitted instantaneously from one location to the next. For any signal a minimum time is always needed to travel a finite distance, since the maximum speed allowed is the speed of light.

## Thought Experiments

*Thought experiments are exercises of critical thinking performed for the sake of testing an element of truth or for revealing a secret of nature.* They are the equivalent, in intuition, of laboratory experiments in which hypotheses are tested under critical physical conditions. Even though reality does not conform to our intuition, Mach saw thought experiments, rigorously performed, as useful tools in describing fundamental aspects of reality. He invented the term *"Gedankenexperiment."* Some typical examples are described below, taken from Sorensen (1991).

### GALILEO'S THOUGHT EXPERIMENT ON ARISTOTLE'S THEORY OF MOTION

According to Aristotle, the larger the mass of an object, the faster it falls in the gravitational field of the earth. According to Galileo, the acceleration of gravity is independent of the mass of an object. That is, a feather falls in a vacuum as fast as a piece of lead.

In his argument against Aristotle, Galileo suggested a simple thought experiment, combining Aristotle's principle that *the heavier the body, the faster it falls,* with a second principle of Aristotelian mechanics which states that *a faster body is slowed in its motion when it becomes attached to a slower one.*

To illustrate the flaws in these principles, Galileo suggested to consider what happens when a big stone gets attached to a pebble in the middle of a free fall. *According to principle 1, the heavier, composite object must accelerate. According to principle 2, it must slow down.*

## A THOUGHT EXPERIMENT REVEALING MACH'S PRINCIPLE

Newton believed that all motion is not relative, but that examples of **absolute motion** can be found. As an example, he described the rotation of a bucket filled with water. When the water in the bucket is at rest, its surface is flat; when it is spinning, its surface is concave. Since this effect cannot be explained by the *relative* motion between the bucket and the water, Newton thought that this was a case of **absolute motion in absolute space.**

Relative motion between bucket and water cannot explain the observed phenomena because, at the beginning of the experiment, at time-1, when the bucket is not spinning and the water not in motion relative to the bucket, its surface is flat. At a later time, t-2, when the bucket begins to spin, for a short while the water remains at rest with respect to the laboratory and is in motion with respect to the bucket, and its surface is flat. But then, after some time at t-3, the water spins with the bucket and its surface bends. If, at t-4, the bucket is abruptly brought to a halt, the water continues to spin for a little while with its surface concave. At t-1 and t-3 the water is at rest relative to the bucket, and yet its surface is flat at one time and concave at the other. At t-2 and t-4 the water is in motion relative to the bucket and yet the shape of the surface is flat at one time and concave at the other. Thus, the shape of the surface of the water cannot be explained in terms of its motion relative to the bucket. Therefore, Newton concluded, it is a case of absolute motion, explained only relative to absolute space.

Mach found this argument particularly objectionable and suggested to replace the concept of *absolute motion* by the concept of *motion relative to all the other masses in the universe.* The principle, later termed **"Mach's principle"** by Einstein, implies that the inertia of a mass vanishes when all the other masses in the universe are removed.

As set forth by Einstein in 1955: *"Mach was led to make the attempt to eliminate space as an active cause in the system of mechanics. According to him, a material particle does not move in unaccelerated motion relatively to space, but relatively to the centre of all the other masses in the universe; in this way the series of causes of mechanical phenomena was closed, in contrast to the mechanics of Newton and Galileo."*

Mach arrived at his principle through a simple thought experiment: *Hold the bucket steady,* he suggested, *and rotate the fixed stars, and then see what happens to the shape of the water surface in the pail.* Where Newton was convinced that *"absolute space, in its own nature, without relation to anything external, remains always similar and immovable,"* Mach objected that *"no one is competent to predicate things about absolute space and absolute motion: they are pure things of thought, pure mental constructs, that cannot be produced in experience."* (For additional details, see Bradley, 1971.)

# Appendix 3

# SOME ASPECTS OF CAUSALITY

We speak of **causality** when the relation between two events is one of cause and effect; that is, when two events are ***necessarily connected.*** When A, then B. If such a relationship exists, *a future event can be predicted with certainty, exactly.* For example, force equals mass times acceleration (F = ma). When a force acts on a mass, it will change its state of motion. Force is the cause of acceleration.

**Causality is popularly taken to be a principle of nature.** In that case it is often connected with **determinism,** the doctrine that *everything* is determined, because everything has a cause.

If an event cannot be predicted with certainty or exactly, one cannot be sure that it has a cause. Thus, there is a problem with causality as a principle of nature, in that ordinary physical events—such as those involving measurements of the temperature, volume, or pressure of a gas; the strength or voltage of an electric current; the frequency or amplitude of a wave; the position, mass, or velocity of a particle—*cannot ever be predicted exactly.* This is so because the results of measurements of such variables are to some extent unavoidably inaccurate. For all continuous variables, even of the simplest type, a level of precision can be selected at which repeated measurements will disagree. When the length of a desk is measured with high precision, a different result is obtained each time. In this sense, no object has a definite length, no tower a definite height, no planet an exactly repeated orbit.

To deal with this predicament, **Planck** proposed that the task of science is to construct a *world image,* a model of reality. Each *observable property* of the physical reality is represented by a *variable parameter* in the model reality. Specific values of the model parameter correspond to *results of measurements* made in the physical reality. To make use of the model, the starting values of its parameters at one time are taken from the external world, and the model laws are then allowed to take their course. At a later time a translation is made from the one world to the other by comparing the value of a model parameter with the outcome of an experiment. If the two agree—*approximately*—a later event has been connected with an earlier one. Strict determinism applies to the variables of the world image, the constructs, but not to those of the physical reality, the observables. The period of oscillation of a pendulum is given exactly by Galileo's formula, but not by any timing. The triangles of geometry are exactly triangular—or as Plato put it, the ideas of triangles are perfect—real triangles never are. All entities of the world of the senses are approximate; those of the model world are exact. We say that translation inaccuracies arise whenever the value of a parameter of the image is compared with the outcome of a measurement.

The handling of the inaccuracies is a matter of disposition. Some people, Planck calls them **determinists,** disregard the deviations, searching for order in chaos. Others, **indeterminists,** are inspired to argue that *acausality* is the fundamental principle of nature, which is all chance, randomness, indeterminacy.

The *gas laws,* connecting pressure, volume, and temperature of a gas, are typically causal laws. Raise the temperature of a gas at constant volume, and pressure will rise. Raise pressure at constant temperature, and volume will decrease. These laws are inexorable, strictly connecting one state with another. At the same time, if one considers the microscopic origins of these laws, it is found that pressure, volume, and temperature are averages of fluctuating molecular properties. Temperature is a measure of average molecular kinetic energy. Pressure is average momentum transfer, caused by the irregular impact of many single molecules on the walls of a container, and density is average number of molecules per unit volume. Considering an infinitesimal volume, at any one moment it may contain no molecule or several of them. Any tiny area of the container wall may not be struck by any molecule for some time. Then, all of a sudden, it may be struck by several at once. In agreement with this nature of a gas, the origin of the word *gas* is traced by some to the Dutch word *chaos,* both very similar in Dutch pronunciation. The conclusion can be that *strictly causally determined quantities, such as pressure, volume, and temperature, are really nothing but statistical averages of highly fluctuating microscopic properties.* Or, conversely, that *even where phenomena seem random, nevertheless causality rules.*

Quantum mechanics aggravated the translation inaccuracies from the model world to the real world for two reasons: Due to Heisenberg's uncertainty (see Appendix 11) not all physical observables can be known exactly at the same time. In addition (as shown below) individual quantum events cannot be predicted with certainty. For example, when a radioactive compound decays, it does so with a constant half-life. After the time of its half-life, one half of the starting amount will have decayed—half a pound when the starting material was a pound, half a ton when it was a ton. Statistically, among a large number of radioactive atoms, half will decay within a well-determined half-life. However, when a particular atom is singled out, nothing is known about the time of its decay—it may happen in a fraction of a second, or not in thousands of years. This is an example of a typical quantum event. Occurrences of single quantum events cannot be predicted precisely, indeterminists say, because nature is acausal. Not so, determinists say: it is not that nature is acausal, but the information available is not complete.

The third possible view of causality is the **positivist** view, postulating that neither causality nor acausality is a principle of nature. Rather, both are metaphysical concepts which have no place in science. In true science, positivists say, one should stick to the data, to the positive, and avoid references to such poetry as tales of causality.

# SCIENCE AND THE HUMANITIES

It has often been said that the physical sciences and the humanities have nothing in common, but represent two different cultures which are totally disconnected. However, the rule of aesthetic principles both in the arts and the sciences shows that this view is not complete and connections exist which can be the basis for fruitful interactions and creative developments. The following concordance of dates is evidence that all cultural activities are in some way connected.

In 1900 Planck founded quantum mechanics; Freud founded psychoanalysis. In 1903 Ford founded his Motor Company, and the Wright brothers succeeded with the first motor flight. In 1905 Einstein published his paper on special relativity and the first show of modern art (Fauvism) was held in Paris. The birth year of Cubism was 1907; of abstract painting, 1910. In 1910 Schönberg wrote the first atonal music. Kafka's *Short Stories* were published in 1913; James Joyce's *Dubliners,* in 1914. The same year also saw the beginning of the First World War, which, in 1917, led to the Russian Revolution.

Each of these dates marks a revolution in its respective field. Together with other seminal developments that could be listed in this context, these revolutions occurred in the fine arts, drama, music, politics, economics, science, and engineering. Within a few decades, around 1900, Europe saw a drastic change in the general perception of reality. A matter of great importance is that *all aspects of culture were involved in this process.* We have to suspect that the concordance is not accidental, but that simultaneity of creation indicates similarity of content.

Among the common features we find that, in the physical sciences and in the arts, the development was characterized by a certain *loss of realism,* a *loss of perceptive models,* and a *break with the classical tradition.*

For example, before 1900 paintings were typically realistic, related to the plane of the human senses, to visual models of reality. The same is true for classical physics in that it was based on simple visual concepts, like waves, particles, or fields. Objects studied could be experienced directly. Objective and absolute principles were applied, such as causality, determinism, objectivity of reality, all of which were thought to be fundamentally rooted in nature and allowed for a conceptually simple, mechanistic view of the world.

All of a sudden a part of reality was discovered—such as the subconscious and invisible world of the elementary particles—that lies *beyond the level of the human senses* and could no longer be described in terms of simple visual models. In physics, abstract mathematical symbols began to be used to represent the newly discovered part of reality. At the same time, in the arts, non-perspective abstract painting became *evocative* (Haftmann, 1965), replacing *reproductive* painting.

Another common motif in the changes that occurred around 1900 involves the dissolution of certain characteristic constants on which conventional thinking had centered in various disciplines. In biology, the belief in the constancy of life forms was replaced by the concept of evolution. In physics, the atoms, once believed indivisible, were disintegrated. In astronomy, the immutable heavens were replaced by the evolutionary universe. In psychology, the structure of the subconscious was discovered, replacing the concept of the homogeneous soul. In painting, the vanishing point of perspective was lost in aperspective paintings, or those with multiple perspectives. In music the keynote of classical compositions was abandoned in atonal music. At the same time, there was a general loss of social constancy, and monolithic empires were destroyed that had existed for centuries.

At the end of the nineteenth century, it was believed that the basic problems of physics had essentially been solved. *The power of humankind seemed unlimited.* There were moves in the U.S. Congress to close the federal patent office because, with the same ability for deep insight that prevails today, the representatives had determined that everything worthwhile had been invented. In contrast, the symbols of the science of our time are *limitation, uncertainty,* and *impotence.* We are not able to attain the speed of light or the absolute zero of temperature. Matter can never be at rest. Dynamic variables occur in pairs; when one is known exactly, the other one is unknowable. Spontaneous processes can occur only in systems which are not at equilibrium; that is, by definition, they are unstable. Thus, the most characteristic aspect of life, its spontaneity, is at the same time a sign of its instability.

We can turn again to the connections between matters scientific and ethical: *A person who believes it is possible to be in complete control, to have full power and unquestionable knowledge, has a different outlook on life than a person who is aware of the limitations and handicaps of human existence.*

Intellectual coherence and concordance of dates is also found in other historic periods; for example, the European Renaissance. Within a century, on a slower time scale than in our time, important revolutions occurred in connection with the following men: Johannes Gutenberg (1398–1468, printing); Sandro Boticelli (1444–1510, art); Christopher Columbus (1446–1506, geography); Lorenzo de Medici (1449–1506, politics); Leonardo da Vinci (1452–1519, art); Desiderius Erasmus (1466–1536, theology and education); Vasco da Gama (1469–1524, geography and colonialism); Jakob Fugger (1459–1525, multinational trade and finance); Nicolai Macchiavelli (1469–1527, political science); Albrecht Dürer (1471–1528, art); Nicolaus Copernicus (1473–1543, astronomy); Michelangelo (1475–1564, art); Martin Luther (1483–1546, religion); Henry VIII (1491–1547, state-craft); Andreas Vesalius (1514–1564, anatomy).

# POPPER'S LOGIC OF SCIENCE

Hume's problem—*the principles that we use to establish scientific knowledge cannot be used to establish them*—and the classical problems in general of defining a logically consistent foundation of empirical science represent a serious challenge to the otherwise strict internal consistency of classical science. One of the most convincing attempts to deal with the dilemma is the philosophy of science developed by **Karl Popper** (1935, 1984). Often unpopular with professional philosophers, it has been enthusiastically adopted by many natural scientists. **Magee** has offered an excellent introduction to Popper's views in his book *Philosophy for a Real World*. This chapter is a summary of Magee's examination of Popper's philosophy; his book (Magee, 1985) is strongly recommended for further reading.

At the outset of his book Magee considers the *orthodox view of science,* that science has to do with the *laws of nature.*

There are two types of laws: *prescriptive* (they predicate what should be done and can be broken) and *descriptive* (they inform of what, given certain conditions, can happen). In the orthodox view, the task of science is to search for descriptive laws.

Since the times of **Francis Bacon** (1561–1626), scientists have generally been expected to go about their business by using the **method of induction,** meaning that in regard to a given problem repeated experiments are performed, particular observations are made, data accumulate and are published, and after some time some general features emerge, inspiring general hypotheses; these lead to the formulation of natural laws and the discovery of causal relations.

*The inductive method is often seen as the criterion that allows one to draw a* **line of demarcation** *between science and non-science:* Scientific statements are based on observational evidence, facts. Statements of any other kind—artistic, political, religious, aesthetic—are based on emotions, tradition, speculation, prejudice, habit, authority—not on facts. In this sense scientific statements are supposed to provide sure and certain knowledge. This defines the orthodox view of science.

The problems with this view were pointed out by Hume:

- No number of singular events allows for a general statement. Laws of physics do not logically follow from instances observed.

- As for causality, repeated observations of "if A, then B" make us infer a necessary connection, but this is Psycho-Logic, not Logic.

- Confirmation of scientific laws in the past is no guarantee that they will hold in the future. No future events have been observed.

- The contrary of matters of fact is always possible.

We conclude: *scientific laws have no rationally secure foundation,* not in logic, not in experience. Induction cannot establish induction.

At this point it is important to realize that, once a small number of basic principles—such as causality and induction—are accepted as valid on faith, everything else follows. The problem is that nothing in science should be accepted on faith and not a single departure from scientific principles can be allowed.

From this one problem, skeptics, irrationalists, mystics, and religious dogmatists derive alleged support for their positions. Therefore, says Popper, **reject the orthodox view of science.**

For the further containment of the problem, Popper turns to the **logical asymmetry** that exists between verification and falsification: *Empirical generalizations, though not verifiable, are falsifiable.* Scientific laws, albeit not provable, are testable. They can be tested by systematic attempts to refute them.

As an example, take the assertion *"All swans are white."* No large number of observations can verify this statement, but a single black swan can falsify it. This sounds simple, but we have to distinguish between the logic of this asymmetry and the implied methodology.

The **logic** of the situation is simple: if a single black swan is observed, the statement made above is false. That is, in logic, any scientific law is conclusively falsifiable, even though not verifiable.

In terms of methodology the matter is not that easy. One can always (a) doubt a given observation or (b) make ad hoc changes to save a given law. *Thus, what seemed to be a black swan, maybe was not a swan, but some other bird. Or one decides, since it was black, not to call it a swan.* Popper's advice for the practice of science is therefore this:

(1) Do not try to evade refutation systematically.

(2) Do not use ad hoc hypotheses.

(3) Do not reject inconvenient experimental results.

(4) Formulate theories so that they are falsifiable. (I often recommend to my students, tongue-in-cheek, *try to make mistakes.*)

As an example of a chain of ad hoc modifications made to save a theory, take the thesis that "water boils at 100°C." No number of confirming instances will prove this statement, but it can be falsified by the observation that water does not boil at 100°C in a closed vessel. An ad hoc hypothesis to save the original statement: "Water boils at 100°C *in all open vessels.*" But water does not boil in all open vessels at 100°C at high altitudes. An ad hoc hypothesis: "Water boils at 100°C in open vessels *at sea level.*" Instead of making ad hoc additions, one might better have asked, "Why does it not boil at 100°C in closed vessels or at high altitudes?" This would have offered a chance of producing a richer, better theory.

From each refutation of a scientific law eventually springs a new theory. Knowledge is provisional. We can never prove that what we know is true; it can always turn out to be false. This is how science advances. One theory replaces another.

Growth of knowledge proceeds from problems and our continuing attempts to solve them. Attempts to find a solution go beyond existing data. A *leap of the imagination* is involved. This must be done not by inductive reasoning, the slow ingestion of existing data, but by bold and creative imagination. Science is not a body of established facts. The orthodox attitude that *science should prove a theory* is false. Rather, *science should justify preference for one theory over another.*

The concept of *"ultimate truth"* is a metaphysical notion. Like the concept of "exact length" or "uninterrupted happiness" it does not exist. All measurements have an appended uncertainty; all feelings will, at least temporarily, induce their opposite.

The demise in this way of the orthodox view of science raises the question as to *where then does a scientific theory, like Newton's, come from?* Popper's answer: it comes from Newton.

The meaning of this answer is that scientific creation and artistic creation are processes of the same kind. Just consider the similarities between modern art and modern science, described in Appendix 3. Induction claims that it says something about how a theory is derived. But that question is not even important. Important is this: Is the theory a good theory? Is it internally consistent? How does it compare with other theories? How a theory has been derived has no bearing on its value. We must distinguish between the logic of a theory and the psychology of its creators. Induction is a psychological process, not a logical one. For the result, a theory, it does not matter how it was derived. Perhaps by dreams, flashes of insight, the reading of tea leaves. There is no illegitimate way of deriving a theory. The most common mechanism is perhaps based on modifications of existing theories. But in general, in Magee's words (1985): *"There is no logical way for having new ideas. Every discovery contains an irrational element."*

There are various reasons why knowledge is not founded on induction: (1) observations and experiments do not give rise to a theory, but are derived from it; (2) formulations of a theory go beyond existing data; (3) many theories fit a limited data set.

It is an important aspect of this process that some theory is always presupposed when observations are made. That question is difficult to answer: Which comes first? The hypothesis or the observation? Answer: an earlier hypothesis. Thus, there is an infinite regress from our current level of knowledge to more and more primitive theories; to unconscious, inborn expectations.

At this point the question returns as to *what then is the criterion of demarcation that separates science from other intellectual activities?* Traditionally science differs by the inductive method and in that it provides statements about reality that express absolute truth. This view is in error. Absolutely certain statements often are trivial: *"It will rain." "Milk is white."* The probability of these statements to be true is 100 percent; the information content is zero.

Instead of making true statements, one should try to make statements that are falsifiable. *"Bolder theories are better ones. The wrong view of science betrays itself in the craving to be right"* (Magee, 1985). Hypotheses are there to be falsified. The realization that science is not striving for certainty has a liberating effect. It is not a grave misdemeanor to conceive a theory that eventually is falsified, because that is the fate of all theories.

Thus, scientific theories must

solve some problem;
contain predecessor theories;
be falsifiable;
agree with all the data.

Marxism, psychoanalysis, astrology, in Popper's view, are not testable, because everything can be explained by them. They are not science, but myths. That does not mean they are non-sense. The difference between science and non-science is not that between sense and non-sense. In fact, all myths contain inspiring suggestions and science was derived from them. Empirical concepts were often formulated as myths before the observations were made.

This view differs from that of the **Logical Positivists** who believed that there are two types of propositions: (a) statements in math and logic that can be established without reference to the external world, and (b) statements about the empirical world; they are established by observation.

To the Logical Positivists: *Every statement which is not either a formal proposition in logic or math or not empirically verifiable is meaningless.*

For Popper, that excludes all science; it renders metaphysics, myths, and religion meaningless. Yet all science evolved out of them. As to metaphysical theories, not only can they be meaningful, they may even be true. If untestable, they cannot be science, but they can still be discussed.

Nevertheless, for positivists, verifiability was taken as the criterion of demarcation between science and non-science, meaningful and meaningless statements about the world. **Ludwig Wittgenstein** (1889–1951): *"The world is everything that is the case."* And, *"Things you can't say anything about, you should not talk about."*

**To summarize, in the traditional view scientific activity is composed of a sequence of steps consisting of (1) observation, (2) inductive generalization, (3) formulation of hypothesis, (4) verification, and (5) recording of established knowledge. Instead, Popper's sequence of steps proceeds from (1) discovery of a problem with an existing theory, (2) deduction of testable propositions, (3) testing, (4) establishing preference between competing theories, and (5) returning to 1.**

Popper's circular sequence implies regression to inborn expectations. The origin of theory used in (1) was a previous cycle. In this way, all organisms are constantly engaged in such problem solving. Evolution by natural selection is a process of error elimination.

As Popper puts it, *"The tentative solutions which animals and plants incorporate into their anatomy and behavior are biological analogues of theories. Like theories, organs and their functions are tentative adaptations to the world we live in . . . In the world described by physics, nothing new can happen that is truly new. A new engine may be invented, but we can always analyze it as a re-arrangement of elements which are anything but new. Newness in physics is newness of combinations. Biological newness is an intrinsic sort of newness. A novelty that cannot causally or rationally be explained."*

*Biological newness is intrinsic.* Emergence of reality from potentia. In the development of an organism, every individual passes through the stages the species passed through in the course of evolution. The same is true for the developing mind. Each individual must repeat the steps humankind has taken to enlightenment and will eventually stop at a certain level, if proceeding to the next does not seem propitious. Thus, the evolutionary sequence is tested, again and again, providing the model for science. In this way our origins, nature, and mental structures appear as integrated elements of one grand web.

**The search for truth, then, is a never-ending effort. From P1 (initial problem) to TS (trial solution) to EE (error elimination) to P2 (new problem).**

All observation is theory soaked. At each step, we are emotionally involved. We cannot start anything from scratch, are always indebted to others. The inventor is irrelevant, a servant. The tests are carried on by others. The masters of the great cathedrals are anonymous, while their creations continue to inspire and shine in beauty.

**This is the wonderful world of objective knowledge.**

# SOME PROPERTIES OF WAVES AND PARTICLES

## *Waves*

Waves are characterized by phenomena that only waves can produce, because waves can **add** in a characteristic way. When two waves coming from different sources and directions meet at the same spot in space, they become components of a single wave whose amplitude (the height of a crest) is either larger or smaller than the amplitudes of the components, depending on whether a crest of one wave adds to the crest or a trough of the other. Wave addition of this kind is called the **interference of waves.** It is a characteristic aspect of wave interference that two waves which at one time interpenetrate, interfere, and superpose into a single wave, can reemerge as individual waves unharmed at a later time, resuming their individuality as though nothing had happened and continuing along the original line of propagation.

There is a principle called *Huygens' principle* after its discoverer, the Dutch physicist **Christiaan Huygens**. It states that *every point in a wave can be considered the source of elementary wavelets that spread out equally in all directions.* In a train of waves that propagates in space, the individual wavelets are not seen because they interfere, superimpose, and merge. But when a wave train strikes a barrier with a hole, a single wavelet can be seen to emerge behind it. Since it spreads out equally in all directions, waves bend around the edges of objects that obstruct their paths. This phenomenon is called **diffraction.**

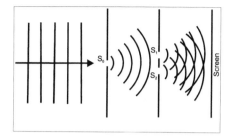

FIGURE 7

*The Generation of a Diffraction Pattern by Two Slits:* When a barrier with a single slit is struck by a wave train (left side of the Figure) it becomes the source of elementary wavelets which spread out in all directions behind the barrier. Here a planar wave train—one whose wave front is a plane—is shown striking a single slit $S_0$. When the wavelets emanating at $S_0$ strike a second barrier, but with two slits in it, $S_1$ and $S_2$, each of the latter becomes a new point source of elementary wavelets which spread out as shown. In this process wavelets sent out from $S_1$ will interfere with those sent out from $S_2$, constructively in some places— where hills of one get to lie on top of hills of the other. If the waves used in the experiment are lightwaves, bright regions— constructive interference—will be seen on the screen to alter- nate with dark regions—destructive interference. The pattern of fringes of alternating bright and dark regions is a diffraction pattern.

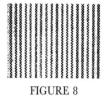

FIGURE 8

A typical interference pattern obtained by the diffraction of lightwaves by a system of slits. The intensity distribution in the pattern is a function of the number of slits.

## YOUNG'S DOUBLE-SLIT EXPERIMENT

At the beginning of the nineteenth century, **Thomas Young** performed diffraction experiments with light, allowing lightwaves to diffract and interfere after they passed an array of slits. For example, when a light beam encounters a barrier with two slits, each of them becomes the source of elementary wavelets which can be seen to emerge behind the screen, superimposing and interfering. Along certain lines from the center of the slits, the waves are exactly in phase (crests are on top of crests, troughs on top of troughs) and reinforce. Along other lines they cancel because they are out of step in such a way that crests of waves from one slit coincide with troughs of waves coming from the other. In the first case, constructive interference enhances brightness. In the second case, destructive interference leads to darkness. Therefore, on a screen behind the two slits a **diffraction pattern** is observed, a pattern of fringes in which regions of darkness alternate with brightness. From the way this pattern is engendered it is clear that its **intensity distribution**— that is, the variation of darkness and brightness—is the result of the interference of waves that originate in both slits, not just in one of them. Specifically, *if the number of slits is changed, a different number of wavelets will interfere, and the diffraction pattern is changed.*

**The example shows how waves are extended in space, not localized. Waves are characterized by their wavelength (the distance between the two consecutive crests or troughs) and frequency—the number of crests or troughs passing a spot in space each second. Wavelength is a "length," not a point, and the passing of crests and troughs is a continuous process, not a discrete corpuscular phenomenon.**

## PARTICLES

In contrast to waves, **particles** are discrete, non-continuous, individual lumps of matter, definite amounts of energy. As in the case of waves, everybody has some experience with such things. Billiard balls are good examples of particles. They are *hard, solid, filling space, exactly localized, isolated, precisely delimited.* They have a well-defined outline, a definite shape.

Interactions of particles, like those of waves, lead to characteristic physical phenomena: In *collisions* one particle may impart energy and momentum to another, and there are characteristic changes in direction and speed of motion. Thus, a typical property of particles is that they can **push.**

**Linear momentum,** symbolized by **p**, is defined as the mass of a particle, symbolized by **m**, multiplied by its velocity, **v**: or **p = mv**. In a mechanical process, the sum of the linear momenta of interacting particles is constant. We say that *in a mechanical process linear momentum is conserved.* When billiard balls collide and move in a certain way, they do so because **p** must be conserved. Similarly, a spinning motion or **angular momentum** is conserved in the interaction of particles. Due to the conservation of angular momentum an object continues in a spinning motion when it is not perturbed, like the earth continues to spin about an imaginary axis through its poles. *It takes a force to change the spinning motion (the angular momentum) or linear motion (the linear momentum) of a moving particle.*

In addition to being characterized by **dynamic variables** (momentum, angular momentum, position, velocity), particles are characterized by the fact that they have mass, m, and thus are subject to **gravity** and **inertia.** *Because of gravity, masses attract; because of inertia, they continue in a state of motion when no force is acting on them.*

It is obvious that the properties of waves and particles are incompatible. Waves are delocalized, can interfere, superpose, interpenetrate, diffract, and some types can propagate in empty space. Particles are localized, inevitably massy, and subject to gravity and inertia. They can push, collide with each other, bounce around, and they never interpenetrate. In view of these differences, we would not expect a single entity to have properties of both particles and of waves.

*Nevertheless, at the turn of the twentieth century, light—electromagnetic radiation—was found to display the characteristics of waves in some experiments and of particles in others. Similarly, elementary particles—lumps of matter like electrons, protons, neutrons—were found to behave like waves in some experiments and like particles in others. The conceptual consequences of this discovery are so enormous that it can be ranked as one of the most important findings in the entire cultural history.*

Light shows the signature of waves in interference and diffraction experiments. The laws of optics, which correctly determine the performance of optical instruments, are based on the wave nature of light. But in addition, light can also show the signature of particles in collision experiments, when individual light particles (called photons) collide with individual electrons, and in doing so obey the laws of mechanics and gravity.

In experiments conducted by **A. H. Compton** in 1923, x-rays—a form of electromagnetic waves like lightwaves—were scattered by various materials, such as graphite. When x-rays pass through a piece of graphite, *they get pushed around* by the electrons in this material and are *scattered;* that is, the direction of propagation is changed. Depending on the change of direction, their wavelength is also changed. In studying this phenomenon, Compton found that he had to use two principles—*the conservation of energy and momentum*—in order to explain the experimental results. Now, momentum, involving the product of mass and velocity, is exclusively a property of particles. Thus, whereas the wave nature of x-rays is revealed in diffraction experiments, in Compton's experiments *x-ray particles* seem to collide with the electrons in graphite and, like in collisions of billiard balls, the electrons fly away in one direction and the x-ray particles in another. The same dual nature has been established for all forms of electromagnetic radiation. We speak of **photons** when we refer to processes in which light particles are involved.

A process involving photons is the **photoelectric effect**. In photoelectric processes, electrons are ejected from a metal surface that is irradiated with light. The energy of the emitted electrons is found to depend on the frequency of the light, confirming a formula, $E = h\nu$, derived by **Max Planck** in 1900, stating that *the energy of electromagnetic radiation* **E,** *is equal to the frequency,* $\nu$ *(the number of waves per second), multiplied by a constant,* **h,** *now called Planck's constant)*. According to classical physics, frequency has nothing to do with the energy of lightwaves. Rather, wave energy depends on intensity, or the square of a wave's amplitude. The greater the amplitude, the greater the energy. In contrast to this, it was found that when very intense light is used to irradiate metal surfaces and the frequency is below a certain threshold, *no* electrons will emerge. At the same time, low intensity radiation with high frequency will lead to the emission of very energetic electrons. The only aspect determined by intensity is the number of electrons emitted: the number is high for high intensity and low for low intensity. High intensity means many photons.

It was actually from Planck's formula, $E = h\nu$, that the notion of photons arose. The reason is that, in a beam of monochromatic radiation (consisting of waves with a single color or frequency, $\nu$), only the exact amount of energy represented by $h\nu$, or integer multiples of it, can be exchanged; not a fraction of $h\nu$, or any non-integer multiples. In this picture, a light beam appears like a stream of coins.

If it consists of, say, a stream of dimes, then only ten cents can be obtained from such a stream, or 20, or 30, and so on, but not 7, or 9, or 92. When the intensity of a light beam drops with increasing distance from the source, the photons in it retain the same size (energy), but the density (number per volume) decreases.

*That the energy of a wave depends on its frequency was a finding in total contradiction to classical physics, and it marks the beginning of the breakdown of classical theory. It is a peculiar notion that makes no sense on the level of conscious experience. It implies, for example, that when you go for a swim in the surf of the ocean, a soft but high-frequency ripple on the water will throw you on the beach with great din and violence in contrast to a tidal wave, whose arrival you can await without any care or concern as long as its frequency is low. In a simple way, $E = h\nu$ was the first indication of the unexpectedly alien nature of the entities underlying the world of our conscious experience.*

Like lightwaves, elementary particles have a dual nature. As particles, electrons and neutrons, for example, have mass, charge, spin, and momentum, and typically collide, push, and are subject to gravity. As waves, they can superpose, interfere, and diffract. For example, electron diffraction is used in electron microscopes, where electron beams are formed and focused like beams of light, and the electron wavelength can be measured. The formula $\lambda = h/p$ shows that the greater the momentum of a particle, p, that is, the faster it moves, the smaller its wavelength, $\lambda$, where h is a Planck's constant. The formula applies, in fact, to all material things in motion.

*The ability of elementary entities to display both the properties of waves and those of particles is* **the wave-particle duality.**

Waves interpenetrate, particles bounce off. Waves are continuous, particles discrete. Waves are extended in space, particles localized. Waves are non-material, mass-less, particles are massy. But never mind how mutually exclusive the properties of waves and particles, **quantum entities** (photons, protons, electrons, molecules) display them all.

**Quantum entities observed act like particles; not observed, like waves.**

# SCHRÖDINGER'S WAVE MECHANICS

The wave properties of elementary particles are among the phenomena that inspired the formulation of *Schrödinger's quantum or wave mechanics,* named after **Erwin Schrödinger** (1887–1961). In this theory, every particle (or system of particles) is represented by a wave function, usually denoted by the Greek letter $\psi$. The wave function is obtained by solving a wave equation—Schrödinger's equation—which is similar to the equations that are used in optics to describe the propagation of lightwaves. To apply the theory to a given system of particles, Schrödinger's equation must be solved for the special conditions of that system. The mathematical properties of the solutions, $\psi$, mirror, in mathematical space, the physical properties of the wave-like states in which systems of particles evolve when they are not observed. The squares of the amplitudes of Schrödinger's wave functions or, as we say, the squared magnitudes $|\psi|^2$, have the meaning of probability density. The description of nature by $\psi$ is complete in that it contains all the probabilities of possible outcomes of measurements on the system that it represents.

The properties of physical systems of particles in real space are characterized by **observables,** *dynamic attributes* such as *position, momentum,* orientation of *spin, energy,* et cetera, whose values are variable, depending on the specific state of a system, and can be determined by measurement. In real space, the values of physical observables are obtained by **experimental operations**. In mathematical space, the equivalent of the physical operations are **mathematical operations** on $\psi$. Thus, mathematical **operators** in $\psi$-space represent the physical observables (dynamic attributes) of real space.

To measure the property of a physical system one has to use the right kind of instrument—that is, one has to perform an operation appropriate for the property in question. It takes a different instrument, or a different operation, to measure, say, the energy of a particle than its angular momentum or position. Exactly the same condition holds in mathematical space: it takes a different mathematical operation to elicit from a given system wave function the information that it contains on energy than it does to elicit information on position or angular momentum. *In Schrödinger's wave mechanics each dynamic variable is assigned its own operator,* or operation. There is an operator for energy, for functions of position, for momentum, et cetera. For each dynamic attribute in real space, a corresponding operator exists in mathematical space.

As it turns out, when a Schrödinger operator works on a wave function, it has the effect of multiplying that function by a constant numerical value. Each Schrödinger operator is characterized by such a constant—sometimes a whole spectrum of them. Therefore these constants are called *characteristic values* or **eigenvalues,** a term taken from the early German literature that means the same thing—characteristic value.

To repeat: when a mathematical operator works on a wave function, it engenders a characteristic numerical value. When an experimental operation is performed on a physical system, it engenders the value of an experimental result. This then is one of the main postulates of Schrödinger's mechanics: *the eigenvalues associated with an operator in mathematical space are values of the possible outcomes of experimental measurements (of the property that the mathematical operator represents, and of the system in the state that the wave function represents).* In this sense it is also seen that the wave function has the meaning of a probability function: when a single experiment can have several outcomes, the wave function provides probabilities for each of them to occur.

Postulates are at the foundation of every physical theory and cannot be verified but tested. Therefore, we cannot speculate as to why the seemingly strange postulates of Schrödinger's mechanics should make sense, or how anyone in the world could possibly have conceived of them. First, much of the current interpretation is being made with hindsight; and second, it was Schrödinger's historic achievement, an extraordinarily creative act, to find the expressions for operators and functions that correctly predict the results of measurements. No matter how it was derived, Schrödinger's equation has predicted many results before they were known and to date no case has been found in which it was falsified.

Apart from the utility of the equation, its success suggests that mathematical analysis can be trusted to reveal aspects of reality. Why this should be true is not immediately apparent and is perhaps yet another sign of the mind-like nature of the background of reality.

# THE MEANING OF Ψ

To put the theses presented in this book in perspective, it is useful to consider some additional interpretations of the meaning of the quantum wave functions which are also possible and in contrast to the views adopted by this author. Where some researchers, like Max Born, thought that the quantum probability waves represent merely our knowledge of physical systems and are otherwise not real, others thought that the Heisenberg tendencies are objectively existing entities. This was Heisenberg's view in his later years. In order to illustrate the highly controversial nature of all interpretations of the nature of ψ, three other views which are of historic significance shall be briefly summarized. They are the *Copenhagen Interpretation, Bohm's Pilot Wave Model, and Everett's Many-Worlds Hypothesis.*

The ***Copenhagen interpretation*** is based on the philosophy of **Niels Bohr** (1934, 1958, 1963). Bohr avoids the metaphysical surprises connected with the Heisenberg ontology by assuming that ψ does *not* represent reality, but represents **what we can know** about reality. To Bohr, the task of science is not to describe the essence of reality, but, in his own words, to *"extend the range of our experience and to reduce it to order. . . . We meet here in a new light the old truth that in our description of nature the purpose is not to disclose the real essence of the phenomena but only to track down, so far as it is possible, relations between the manifold aspects of our experience."* (1934)

*"The sole aim . . . [of the quantum mechanical formalism] is the comprehension of observations obtained under experimental conditions described by simple physical concepts. . . . In this connection I warned especially against phrases, often found in the physical literature, such as 'creating physical attributes to atomic objects by observation' . . . . [T]he appropriate physical interpretation of the symbolic quantum-mechanical formalism amounts only to predictions, of determinate or statistical character, pertaining to individual phenomena appearing under conditions defined by classical physical concepts."* (1958)

Thus, according to the Copenhagen interpretation, the collapse of the wave function by an observation is not a sudden change of *the state of the universe,* but a *sudden change in our knowledge* about some phenomena of physical reality. There is nothing particularly dramatic about it. As pointed out by Stapp (1993), in the Copenhagen interpretation the nature of ψ is not a problem of ontology, but epistemology. Stapp's book contains what is perhaps the most thorough presentation of the Copenhagen interpretation.

In **Bohm's pilot wave model,** which originated with ideas by **Louis De Broglie** and **David Bohm,** the wave-particle duality is rationalized by assuming that waves and particles coexist in unison; that is, the world consists of real particles, as we always thought, but in addition each particle is accompanied by a wave which guides its observable behavior and engenders the wave-like aspects of reality. In Bohm's theory each particle *owns* its attributes in a classical sense, *has* a definite position and momentum, and *is* a particle at all times, but in addition is subject to control by an associated wave, **the pilot wave**. In the diffraction of electrons by slits, for example, the ubiquitous pilot wave constantly inspects the states of the slits and instructs the particle how to proceed. When a change occurs, as when one of the slits is closed, the particle is immediately instructed by its associated wave and forced to act accordingly.

In **Everett's many-worlds interpretation**, named after **Hugh Everett,** the unreduced ψ—a superposition of states—is considered to be real, as in Heisenberg's ontology, but when a measurement is made it does not collapse. Rather, *when a measurement is made, it is postulated that all possibilities become real, that the different possible results will actually all occur, but each one in a different universe. For each possible outcome a separate universe is created in a giant quantum jump, with each separate edition being exactly identical to the others in all properties but the one just measured.* The observer, cloned into many identical observers in the different universes, is unaware of his fate and aware only of one actual result, because his universe is disconnected from all the others that were created by the measurement.

Finally, removed from all interpretations of the meaning of quantum phenomena, there is the **pragmatic** view. To avoid the strangeness of the quantum world, many researchers in the field take the quantum theories in a pragmatic way, refusing to consider any conceptual consequences. *Pragmatism is a doctrine that measures the validity of theories by the practical success of their performance.* Pragmatists stick to facts and mathematics and require successful performance of a theory. Action is the goal of cognition. The importance of concepts lies in their practical consequences.

**William James** (1842–1910) is one of the modern representatives of pragmatism. To James, the importance of a proposition is equal to its usefulness. In pragmatic ethics, standards of conduct are not based on superior moral principles, but rather ethics have a purpose, serving society, and actions are adapted to particular situations. In a pragmatic mood, the television commentator of a game, for example, may praise a "good foul," which is one that is considered to have worked because its perpetrator was not caught by the referee.

Pragmatists want nothing to do with speculations on the nature of reality. Thus, there is kinship with positivism and empiricism (do not speculate about things that are not observed).

The various attempts to interpret the quantum phenomena all have their specific advantages and problems. I have found the Heisenberg ontology to be the most convincing, as supported by much of the recent evidence. For additional details I recommend other texts for further reading, such as the books by Bohm (1980), Davies and Gribbin (1992), Gribbin (1984), Herbert (1988, 1993), Polkinghorne (1985), and Stapp (1993).

# EMPTY ATOMS AS PLATONIC FORMS

## *Schrödinger's Atoms and the Concept of Substance*

Things cannot be like the experience we have of them (**Bertrand Russell**). We do not have immediate experience of things, but only of our interactions with things (**Kant**).

The *shape* of an object perceived is some kind of projection, different to different observers. Its color is affected by surface reflections, our eyes select only a few of the frequencies, colors, that bounce off any surface. There is no reason to believe that objects are in any way like the photons that are bouncing off of them.

In physics, *temperature is motion*. But our sensation of hot and cold is a totally different quality than rapid or slow molecular oscillation. The sensation of temperature is not inherent in things. I have not heard a person exclaim, "ouch, too fast for me," dropping a hot potato. In fact, the discovery that heat and mechanical energy are equivalent is one of the more protracted affairs of the history of science.

We *see* that the sun and the planets revolve through the year around our planet. We *say* that the motion observed is not the real motion, but the apparent one.

It is a consequence of Einstein's relativity theory that for any change in the energy of an object there is a corresponding change in mass. If the energy of a system is increased, its mass is increased. Thus, it is perfectly correct to say that a moving person is heavier than the same person at rest; a loaded spring is heavier than the same one relaxed; a hot water bottle is heavier than the same bottle cold. All these statements are true, yet contrary to common sense. In the same way it violates supposedly absolute principles, accepted without restraint in everyday experience, that moving clocks are slow compared to stationary clocks; that matter warps space; that people can age at different rates depending on how deep they are in a gravitational field.

In this and many other ways, the experience of things is not of the true nature of things. This is the problem of *appearance and reality* (Russell, 1978). Truth is not a property of things, but of our sensation of things.

In contrast to the inability to grasp the essence of things directly, or maybe inspired by it, our basic attitude regarding reality has always been marked by a *desire for certainty*. We have an instinct, a craving for certainty. This may be more a Western than an Eastern propensity, but quite generally we are programmed with a desire which is entirely in contrast to the fact that the only things that we can be sure of are our sensations. Only sensations are immediate. The problem is that the inherent properties of things are public, whereas, as Russell (1948) says, "data are private." In Russell's definition (1948, 1978): "sense data" are the things immediately known to us in sensation; "sensation" is the experience of being aware of these things.

Since the data are not like the objects that cause them and the properties of objects are inferred, we can reasonably ask if there is an objective reality that is independent of our senses, and, if there is, what are its inherent properties?

Observable reality has many faces, is ever changing and diversified. In the search for the **enduring, unchanging object** that causes our sensations, the Greeks invented the concept of **substance**: the unchanging element that sustains all changing appearances. From this arose the concept of **matter,** the stuff that all physical (real) objects are made of, solid, indestructible, and lasting. The main properties of stuff were originally postulated as uniformity, indestructibility, and solidity. Etymologically, "matter" has the same roots as "mother." In Latin, matter is *materia;* mother is *mater.*

**Parmenides of Elea** (c. 540 B.C.) connected the concept of matter with the concept of space, claiming that *"to be"* means *"to fill space." "Only what is, is. What is not, cannot be and cannot be thought of."* According to this, there is no vacuum, because it is nothing.

The impossibility of the void has some interesting implications. Since what is moving needs an empty place where it can move to and since there is not empty space, Parmenides also concluded that *there can be no motion.* The same is true for *becoming.* In his poem "on nature" he claimed that *nothing ever changes* (see Russell, 1946). If our senses tell otherwise, they are deceiving. True knowledge is a matter of reason, not experience. The multiplicity of things is a mere illusion. All observations of becoming, decay, and motion are deceptive. The only true being is *"the One,"* infinite, and indivisible. It is present everywhere, a sphere, a substance. It does not change.

**Thales of Miletus** (c. 640 B.C.) thought that *water* was the primeval substance; **Anaximenes** (c. 494 B.C.) chose *air* as the ultimate element of reality; **Heraclitus** (c. 540 B.C.), *fire*. Since the observed reality is qualitatively diversified, and since there is such a multiplicity of observed phenomena, a single fundamental substance did not seem sufficient for explaining everything. Therefore, **Empedokles** (c. 490 B.C., Siciliy) chose four basic roots: *fire, water, air,* and *earth,* and the variety of different objects is then achieved by mixing. And **Anaxagoras** (c. 500 B.C., Ionia), finally, chose an infinite number of fundamental substances, all essentially different, the *seeds* or *germs* of things.

From here it was a small step to **Leucippus** (c. 440 B.C., Miletus) and **Democritus** (c. 430 B.C., Abdera) and the atomic theory that the world consists of ***empty space, non-being,*** and the ***space-filling, being, solid atoms,*** whose main property is to fill space. Atoms do not have emptiness in them. They fill space solid, but between them is empty space. They are **indivisible.** Real objects are composed of atoms. Decay is the dissociation of objects into atoms. Again the connection is between **"being"** and **"non-being"** and **"full"** and **"empty,"** respectively. Democritus' atoms possess position in space, have a definite size, and different arrangements engender the variety of perceptible qualities (color, smell, etc.) in things. The atoms in themselves do not possess the qualities found in ordinary things. Rather, in Democritus' words (see Russell, 1946): *"By convention sour, by convention sweet, by convention colored; in reality nothing but atoms and the void."* **This sets atoms in contrast to the qualities of things present in sensation. Greek philosophy already anticipated: atoms are not like ordinary "real things."**

**Anaximander** (611–549 B.C., Miletus) further developed this aspect. He also thought all things come from a primal substance, but it was not air or water or fire. **It was not like any substance we know.** He thought that the primal stuff was infinite, eternal, ageless, "it encompasses all the world" and the various ordinary substances arise from it. Anaximander called this the *"apeiron,"* indefinite, infinite, constantly creating new worlds; one cannot help but see the close connection to the concept of the Heisenberg state of the universe. To Anaximander, cold and warm, wet and dry separated out of the apeiron. In Heisenberg's ontology, reality separates out of mathematical forms.

In the further evolution of the concept of matter, it became something opposed to mind, as mind has the ability of assuming a state of consciousness, reacting to sensations with willful spontaneity. **René Descartes** (1596–1650) is one of the prime movers of this development in Western thinking, proposing that the world consists of two types of stuff: thinking substance (*res cogitans*) and space-filling substance (*res extensae*).

In contrast to this, **George Berkeley** (1685–1763) decided that there is no such thing as matter. Matter does not exist and things are nothing but ideas in someone's mind. As for the role of the mind, **Leibniz** (1646–1716) claimed that the world consists of **monads**, some kind of atoms, which all are *rudimentary minds.*

Out of this historical background evolved the concept of the atom, seen for thousands of years—as Democritus saw it—as solid, filling space, indivisible.

But then, unexpectedly, at the end of the nineteenth century, the **electron** was discovered in cathode rays and identified as a particle with a definite mass and electric charge, raising the question as to where electrons come from. To solve this puzzle, **J. J. Thomson** (1856–1940) proposed that *atoms were not indivisible but, in fact, were made of smaller building blocks, elementary particles,* electrons among them. This was the origin of Thomson's "raisin pudding" model, in which atoms were uniform spheres of positively charged matter with electrons embedded like raisins in a pudding.

In 1910, **Ernest Rutherford** (1871–1937) performed experiments in which $\alpha$ particles, rays emitted from radioactive substances, were shot at atoms in thin metal foils. In this process, the $\alpha$ particles collided with the atoms and were scattered; that is, they were somewhat deflected from the direction of the incident ray. Such deflections were expected, but in addition, unexpectedly, some of the $\alpha$ particles were found to recoil, bouncing back in a fashion not consistent with the distribution of mass in Thomson's atom model. The phenomenon was as stunning as the recoil of a cannon ball would be from a piece of tissue, and forced the conclusion that *more than 99.9 percent of the mass of an atom* and all its positive charge are concentrated in a tiny nucleus, whose diameter is approximately 100,000 times smaller than the volume occupied by atoms in ordinary materials. Thus, the notion of the **nuclear atom** was conceived, with a nucleus containing practically all of the mass, and electrons moving outside in the surrounding environment. The stunning aspect of Rutherford's nuclear atom model is this: If the average space taken up by an atom is compared with the size of planet earth, the size of the nucleus is that of a baseball field. *Atoms are not filling space solid, but are empty,* with the electrons moving around the nucleus "like a few flies in a cathedral," as Rutherford put it.

Rutherford's nuclear atom posed a serious problem. Due to the strong attraction between positive and negative electric charges, the electron in an atom should spiral into the nucleus and stay there. No mechanism was provided by classical physics that would allow electrons to keep away from their nuclei. A first partially successful attempt to solve the problem was proposed by Bohr. In Bohr's **planetary atom model,** the electrons were postulated to circle around the nuclei restricted to orbits with fixed radii, excluded from the spaces in between, like planets around the Sun.

Bohr's planetary atom triumphantly reproduced the spectra of atomic hydrogen, but failed to do so for heavier elements, since its basic premises were false. Moreover, Bohr's model suffered from conceptual inconsistencies: the laws of classical physics were applied and then, by keeping the electron at a fixed distance from the nucleus, suspended. Also, as we now know, electrons are not miniature satellites with definite trajectories. As Heisenberg (1962) later on pointed out, "No planetary systems following the laws of Newton's mechanics would ever go back to its original configuration after a collision with another such system. But an atom of the element carbon, for instance, will still remain a carbon atom after any collision or interaction in chemical binding."

Greek atoms were indivisible, filling space solid. They spawned a tradition of thinking that extended through Newton into modern science. *"God in the Beginning form'd Matter in solid, massy, hard, impenetrable, moveable Particles."* At the end of the nineteenth century, Thomson's atoms became some positively charged, soft kind of stuff, dotted with electrons. Then, in Rutherford's nuclear model, the atoms essentially turned into pieces of empty space, with nearly all of the mass concentrated in a tiny nucleus. *"The revelation of the void within the atom,"* Eddington commented on Rutherford's discovery

(Eddington, 1930), *"is more disturbing than the revelation of the void in interstellar space or Einstein's space time."* Knock your head against a wall and you are knocking empty space.

To arrive at a more consistent theory of the atom, and bearing the wave-particle duality in mind, Erwin Schrödinger set up a wave equation (see Appendices 7, 8, and 10) to describe the states of electrons in atoms, similar to the equations used in optics to deal with the propagation of light. *In Schrödinger's atoms electrons are standing waves.* The electron waves in Schrödinger's atoms are *quantum waves,* $\psi$, with all the properties of these waves that we have desribed above. They are empty, carry no mass or energy, and represent information on probabilities—in this case, for finding an electron at a position in space relative to its nucleus. For example, in the groundstate of the hydrogen atom (the state with the lowest energy), the most likely spot to find the elecron is at the nucleus, because there the probability amplitude is largest. However, the quantum wave function of this state extends to infinity, with non-zero values even at large distances from the nucleus. Thus, there is a small chance, albeit vanishingly small, that an electron which goes with a hydrogen atom in your body can be found on the moon. In this way the atoms in a human body are not limited to a spot, but reach out into space.

The spectra of atoms—the result of their interactions with light—show that their electrons exist in various states which differ by energy. Bohr thought that these states represented different orbits, in which the electrons circled the nucleus at a different radius. In contrast, in Schrödinger's atoms, *different states are different wave forms,* corresponding to different patterns of probabilities.

The electron wave functions in Schrödinger's atoms are called *orbitals.* The name—invented in the early days of quantum mechanics —derives from the notion that the electrons are not really in orbit about the nuclei, but somewhat so. Orbitals exist in the space surrounding a nucleus and extend, without any definite boundaries, to infinity.

Since the orbitals form the outsides of atoms, they should determine the shapes of atoms. But what shapes do orbitals have? We can plot such functions, obtaining definite forms, but in actuality these forms are invisible. Nevertheless, we say that these wave forms exist, because *the results of their interference are apparent in observable phenomena.* For example, an important type of interference, and the basis of all chemistry, is that between different atoms when they are forming a chemical bond. The results are apparent in the observable properties of molecules; their structures, for example, depend directly on the atomic wave forms by whose interference they are created.

Since the orbitals and probability densities are not visible, what then are the shapes of atoms? The answer is: *atoms have no shapes.* Ordinary things have a shape but the atoms from which they are made do not. If you could crawl into a hydrogen atom, what would you see? Nothing. Once in a while, hello, a tiny dot, perhaps a flash, the electron. Otherwise: empty space. When you look away, the electron is in a state of possibility, not actuality. Herbert (1988) has given an enjoyable description of such a scenario:

> Suppose we could shrink ourselves and our apparatus to nuclear dimensions and enter a hydrogen wave function like a spaceship exploring an unknown solar system. . . . Could we solve the mystery of what the wave function really represents if we were small enough to get inside an electron's proxy wave?
>
> What would we see in there? Nothing special, I believe. The hydrogen atom has one electron. If we are lucky it will make a single, well-localized flash on our screen. That's all. Electrons are always observed to be particles no matter what the size of the measuring apparatus. The quantum paradoxes do not arise because of the relative sizes of atoms and humans (a mere quantitative difference), but because of a more fundamental qualitative difference between the experiences of human and atomic beings. As Bohr and Heisenberg have pointed out, human experience is inevitably classic, but atoms do not exist in a classical manner.
>
> Indeed, we already know what it would be like to live inside an atomic wave function, because we walk around "inside atoms" whenever we go outdoors on a starry night.

The graphs of orbitals found in chemistry textbooks are mathematically computed boundary surfaces, a manner of representation. They are constructs, calculated surfaces of volumes that capture a certain percentage of the probability of finding an electron around a given atom. No such boundaries are seen in real atoms. Nevertheless, most students completing the course are convinced that this is what atoms actually look like. In contrast, *"one cannot speak about atoms in ordinary language,"* Heisenberg said.

Atomic orbitals are often presented as electron clouds, images blurred by the high speed of electrons—presumed to be tiny balls—racing over probability surfaces. However, assuming that an ordinary particle is in actual orbit about a nucleus in this way is not in agreement with the nature of orbitals. In some states, for example, the orbitals are made up of disconnected parts of space which are separated by regions in which the electron must never be. Such orbitals consist, in a way, of a number of closed compartments. An ordinary particle once found in one of these compartments would be caught in a closed cage, surrounded by regions of space in which it must not enter, and would have no way to move to other parts of the orbital. Thus, the electron is not (Herbert, 1988) *"a busy insect that traverses its spatial realm so swiftly that it seems to be everywhere at once."*

The description of Schrödinger's atoms given above is entirely in agreement with what we have said before: No quantum entity is real in the ordinary way when it is not observed. Electrons in atoms are tendencies, have the potential to become real when they interact. Electrons do not spiral about nuclei in corkscrew-type motions. They are standing waves. Unmeasured, they are extended possibility, permeable, wave-like. Measured, they are actual blips on a phosphor screen. An electron does not *move* from closed compartment of an orbital to the next. Rather, one moment it is here, then there.

## SCHRÖDINGER'S ATOMS AS PLATONIC FORMS

In Plato's philosophy *"eidos,"* or *image*, signifies the common essence of different things of the same kind. There are *proto-images* (the *ideas* in Plato's vision of the world) which are truly real and are represented (imaged or made present) by material things.

In Plato's *Timaeus*, the ultimate constituents of air, the atoms, were thought to be octahedra; atoms of fire, tetrahedra; atoms of water, icosahedra; atoms of earth, cubes. In view of Plato's general philosophy, some of his interpreters have taken his description of atoms to mean not that atoms are actual physical bodies with regular shapes but that, rather, they are **mathematical forms, functions, mind-like concepts. In short, what we are now calling wave forms.**

As a young man, Heisenberg was deeply impressed by Plato's atomic theory. Quite frequently ideas of natural philosophy of ancient times become recurring themes in our thinking, but seen in a different light. In Heisenberg's words (1979):

Modern science has accepted from antiquity the idea of a pattern capable of mathematical description, but it carries it out in a different manner. . . . The realm of mathematical forms at the disposal of ancient science was still comparatively limited. They were primarily geometrical forms which were related to natural phenomena. Hence Greek science searched for static patterns and relationships. . . . Modern science has demonstrated that in the real world surrounding us, it is not the geometric forms but the dynamic laws governing movement (coming into being and passing away) which are permanent.

The elementary particles in Plato's Timaeus are not substance but mathematical forms. . . . In modern quantum theory there can be no doubt that the elementary particles will finally also be mathematical forms, but of a much more complicated nature.

The properties of electrons in Schrödinger's atoms are determined by numbers. They are called quantum numbers because they determine a fixed or quantized amount of various properties, such as energy or angular momentum, that the electrons have in these states. This is very similar to the views of Pythagoras, who believed that numbers contained the mystery of physical reality, and the harmony of the cosmos rested on number ratios. This same idea reemerged in the Romanesque architecture of the eleventh and twelfth centuries in Europe, when every church building was understood as an image of a world whose order was thought encoded in numbers, so a symbolism of numbers was used to design a church and to represent the order of the universe.

# Appendix 10

# SOME ASPECTS OF THE NATURE OF QUANTUM STATES

In Schrödinger's wave mechanics a wave function, $\psi$, represents a physical system; for example, the electrons around the nucleus of a given atom. *Mathematical operations* on $\psi$ generate numerical values, *characteristic values*, which are postulated to correspond to the results of *physical operations;* that is, measurements on the real system. Thus, mathematical operators represent the *dynamic attributes* of physical systems, such as momentum, position, energy, angular momentum, et cetera, each of which is assigned its specific operator. The characteristic value that each operator engenders by operating on $\psi$ depends on the functional form of the latter. *The different states of a system are determined by different wave functions, different wave forms.* When an atom undergoes a transition from one state to another, it changes its probability pattern from one wave form to another. As an example, consider Schrödinger's model of electrons in atoms.

Schrödinger's equation for an electron moving in the field of an atomic nucleus is a differential equation. When it is solved, wave functions, $\psi$, are obtained which depend on integers, *quantum numbers,* usually denoted by $n$, $l$, and $m$. The possible values of these numbers are $n = 1, 2, 3, \ldots$ and higher; $l = 1, 2, \ldots$ up to $n - 1$; and $m = 0, \pm 1$, $\pm 2, \ldots$ up to $\pm l$. Each allowed combination of the quantum numbers defines a different state of the electron with a different form of its state function or wave function, $\psi_{n,l,m}(x,y,z)$. In each state, the electron has fixed or quantized values for energy, angular momentum, and the component of angular momentum along an axis. We say that energy is *quantized* (determined by $n$), and angular momentum (determined by $l$) and its component along an axis (determined by $m$) are *quantized.*

The one-electron wave functions $\psi_{n,l,m}(x,y,z)$ are called *orbitals.* They exist in the space surrounding a nucleus and extend, without any fixed boundaries, to infinity. The meaning of the $\psi_{n,l,m}(x,y,z)$ is the same as before: they are probability amplitudes. Thus, the electrons around atomic nuclei are standing waves—*quantum amplitudes*. The squared magnitudes, $|\psi_{n,l,m}(x,y,z)|^2$, provide the probabilities to find an electron in space.

Electronic states in Schrödinger's atoms are good models to illustrate what quantum mechanics understands by a **state**. Each state is defined by the set of quantum numbers in $\psi_{n,l,m}$. All particles in the same state are postulated to be exactly *identical*. This is a remarkable characteristic, because no two classical objects that we know are ever exactly alike. At the same time, even though alike by nature, two particles in the same state need not behave in the same way. For example, excited states in atoms (those with more energy than the ground state) have a relatively short lifetime, typically picoseconds or less. When all hydrogen atoms in a sample are, say, in the excited state $\psi_{5,1,0}$, then by the definition of quantum mechanics *all are exactly identical*. However, in this ensemble of identical individuals, some may undergo a transition back to the ground state within picoseconds, others much later, perhaps after a thousand or million times longer interval, while others may not go to the ground state at all, but first to some intermediate energy level.

In considering the question as to what causes *identical quantum objects to behave in a different way*, the answer is that there are no discernible causes. In this context Herbert (1988) coined the terms *"Quantum Ignorance"* in contrast to *"Classical Ignorance."* When quantum ignorant, we are ignorant of effective causes because there are none. When classically ignorant, we are ignorant of effective causes because the situation is too complex and information incomplete on factors which, for all we know, do exist. Exact causes exist, for example, of weather, but are unknown, because too many variables are impossible to know. The outcome of tossing an ordinary coin, heads or tails, is not predictable, because details of momentum transfer, position in space, et cetera—that is, details of properties which actually exist and are possessed by the coin—are not known with sufficient precision. In contrast, the outcome of tossing a quantum coin is not predictable, because there is nothing there to be known. *Parmenides: Nothing happens in quantum reality.*

# HEISENBERG'S UNCERTAINTY PRINCIPLE

## *Transitory Species at the Foundation of Continuous Identity: Chaos at the Foundation of Order*

One of the characteristic aspects of the ordinary objects of our conscious experience is the fact that their physical attributes seem innate, inexorably a part of them. The table I write on exists, because it **owns** its position in space; it also **owns** its linear momentum or other dynamic (variable) attributes. In addition, it is a characteristic aspect that any combination of physical attributes can seemingly be measured with some precision, subject only to the limitations of the measuring device. In contrast, quantum particles are different. Their dynamic attributes are conjugated in pairs. If one of them is known exactly, the other one is unknowable.

Consider a particle residing on an axis in space, say the *x*-axis. When its position is measured repeatedly, to some extent the results will vary. In general, all measurements are affected by *experimental error*, afflicted with *experimental uncertainty*. We characterize the differences that may occur between various measurements by the Greek letter *delta*, $\Delta$. When measurements of position, *x*, are made, we say that the *uncertainty in position is* $\Delta x$. For example, if the particle was found at $x = 1.0$ cm within an accuracy of plus or minus $0.01$ cm, then $\Delta x = 0.01$ cm. Similarly, when momentum, $p_x$, is measured along the same axis, it will be known with a degree of uncertainty designated by $\Delta p_x$.

Heisenberg's uncertainty principle states that the dynamic variables of a particle come in pairs which are connected in such a way that the uncertainty in one multiplied by the uncertainty in the other cannot yield a value that is smaller than Planck's constant, h, divided by $4\pi$. Position and momentum are such a *conjugate pair*. When the uncertainty in position, $\Delta x$, is multiplied by the uncertainty in momentum, $\Delta p_x$, the result cannot be smaller than $h/4\pi$, meaning that to the extent we are sure of the value of one we cannot know the value of the other. We write $\Delta x \Delta p_x \geq h/4\pi$.

Heisenberg derived the uncertainty principle by a thought experiment that he performed using a microscope to determine position and momentum of a particle. We are not concerned here with the technicalities of the matter, but will consider some specific situations to illustrate its meaning. Say that, for example, the position of a particle is known exactly. This means that the uncertainty is zero; $\Delta x = 0$. In that case, in order to satisfy the uncertainty principle, $\Delta p_x$ must be infinitely large, because only then the product with $\Delta x = 0$ can yield a value that is greater than $h/4\pi$. When the uncertainty in momentum is infinitely large, *nothing is known about the momentum.* When position is known exactly, momentum is unknowable. And vice versa, when the momentum is known exactly, nothing is known about the particle's whereabouts.

The relationship is exactly what the name says; that is, *a matter of principle.* It has nothing to do with the precision of measuring devices or the skills of experimenters. If an instrument existed that could determine position without any uncertainty and if, further, a second instrument could determine momentum without any uncertainty, Heisenberg's principle says that the two instruments could not do their job at the same time. *The two variables can never be known exactly at the same time, ever.*

All kinds of situations were designed in attempts to falsify the uncertainty principle. None of them has succeeded, because the relation is fundamentally connected with the wave-particle duality. For example, the solution to Schrödinger's equation for a free particle moving in some direction in space, say x, is a wave $\psi(x)$ with an exactly defined wavelength, $\lambda$. When $\lambda$ is known exactly, momentum is also known exactly, since the two are related by the equation $p = h/\lambda$. When momentum is defined exactly, its uncertainty is zero and we say that the particle is in a definite momentum state.

To test the uncertainty principle we will now consider the position of the particle in a definite momentum state. As usual, information about position is given by squared quantum amplitude $|\psi(x)|^2$. For a free particle the wave function $\psi(x)$ is such that the probability to find it is *the same everywhere in the universe.* Thus, the position of a free particle in a definite momentum state is totally indefinite. When momentum is known exactly, position does not exist.

At this point we take note again that *constant probability to be anywhere* cannot be ascribed to an ordinary thing that is somewhere. If it is somewhere, the probability at that point is 100 percent ($|\psi(x)|^2 = 1$) and zero everywhere else in the universe. This is *fundamentally* different from saying that the probability is *not zero* anywhere. The assumption of constant probability is not the same, either, as merely *not knowing* where the thing is, but expecting it is somewhere. If it *acts* like it is everywhere, it does not have a definite position. Not, position is not *known;* rather, it *has* none; it is created by a measurement.

The *non-locality* of quantum phenomena is also apparent at this point. When $|\psi(x)|^2$ has a constant and non-zero value throughout the entire universe, but at one time I happen to find the particle in my laboratory, then, *instantly,* $|\psi(x)|^2 = 1$ in the laboratory and $|\psi(x)|^2 = 0$ everywhere else. *Thus, something I do here now, in that same instant affects the probability web all over the universe.*

When an electron is launched from a pointed filament to a distant detector, the line between the source and location of impact defines its momentum fairly precisely. If, in this setup, the position of the electron at one point along the line would also be determined with some precision, the uncertainty principle could be violated. One might attempt such a deed by inserting a screen with a hole between filament and detector, through which the electron must pass. At the moment when it is in the hole, the uncertainty in its position is not greater than the diameter of the opening. By making the opening increasingly small, the position of the electron is determined with decreasing uncertainty. However, when the opening is made small, all of a sudden diffraction occurs, the electron acts as a wave that spreads out in all directions behind the diaphragm, and any information on its momentum is lost.

In other thought experiments aimed at violating the uncertainty principle, the properties of crystals were invoked. In a crystal the positions of constituent particles can be determined precisely because they are situated at lattice points whose coordinates are part of an extended, repetitive array. In contrast, the momenta of atoms or molecules in crystals are not so precisely known because they are constantly oscillating at random about the lattice points. If that motion could ever be stopped, the momenta would also be precisely determined—being exactly zero—and Heisenberg's principle for atoms in crystals violated.

Motion in materials is related to heat. Add heat to a crystal, and the motion of its molecules about the lattice points will become more energetic. Take away heat, and the motion slows down. Thus, the idea arose that Heisenberg's principle might be falsifiable by taking away *all* the thermal energy of a crystal, cooling it down to the absolute zero of temperature—*zero degree Kelvin,* as it is called—

stopping all atomic motion. As it turned out, molecular motion does not stop at the absolute zero of temperature and Heisenberg's principle is not violated. Even at absolute zero particles are in motion with a minimum amount of energy, *the zero-point energy*, that cannot be removed from them. *"Rest,"* as Max Born once said, *"is a word for something that does not exist."*

## TUNNELING

The uncertainty principle is related to a curious phenomenon known as *quantum mechanical tunneling.* In tunneling, quantum particles penetrate potential barriers in a way which is classically forbidden, because the total energy of the particle in the barrier is less than its potential energy—which is impossible to ordinary things.

When a car, engine idling, coasts downhill into a valley and uphill on the other side, it will run up to a level permitted by the energy gained in running downhill. If its energy is too low to coast over the top, then at some point up the hill it will stop and return downward, in the opposite direction, then go back and forth until it comes to a stop.

Quantum particles are not allowed to act in this way because stopping and returning implies that, at one point, the momentum is identically zero without any uncertainty. Since this happens at a location, the *turning point,* that is exactly determined by the energy, the uncertainties in both position and momentum are zero, and Heisenberg's principle is violated.

To avoid this predicament, quantum particles oscillating in a potential well—a valley of potential energy between two hills or barriers—*must not have definite turning points.* This condition is obeyed if the probability to find the particle does not stop abruptly at a barrier, equivalent to its bouncing off; instead, the wave function penetrates into and through the mountain and out the other side. In other words, even though its energy is too low to climb over the top, there is a finite probability that the particle will pop out on the other side, *tunneling through.* The strange thing about a quantum particle is that its tunneling occurs where no tunnel has been dug.

If you are locked in a room, you can break through the wall if you assault it with enough violence; that is, if your energy is high enough to overcome this barrier. Your quantum person can achieve the escape more gently, by simply tunneling through the door. When a key is accidentally locked inside a room, its wave function will tunnel through the walls and into the hallway. By working with her wave function collapsor on the part of ψ that sticks out in the hallway, the quantum locksmith will make the key come through the door.

In order to avoid the quantum strangeness, the uncertainty principle is often explained (Herbert, 1988) by stating that measuring one of a pair of conjugate variables *perturbs* the other one in an unknown way. The perturbation model is then thought to cause uncertainty *merely for our information* about attributes that otherwise have precise, actual values all the time. In contrast to this perturbation model is the view that *if one of a pair of conjugate variables is known exactly, the other one does not exist*. According to the latter, ordinary things are made up from components which are different in essence. Elementary particles **are real**, but real in a different way. In contrast, the perturbation model considers the world of elementary particles as a miniaturized edition of the macroscopic world of our consciousness. Quantitatively less, but in essence exactly the same. Heisenberg (1962) defended the opposite view: *"If one wants to give an accurate description of an elementary particle, . . . one sees that not even the quality of being belongs to what is described. It is a possibility for being or a tendency for being."*

To have no particular position, or to be *not in a definite state of motion*, is so counterintuitive because there is no equivalent in the experience of self. "I" am always somewhere. "I" am always in a definite state of rest or moving one way, not in several directions at once. Of course, knowing where I am exactly is the reason for not knowing where I will be ten years from now. People who know the latter, have no idea of their whereabouts. Since we would immediately tell them, rascals that we are, they are forever hidden from us.

## *Uncertainty of Energy and Time: No Observable Continuous Existence*

A second version of Heisenberg's uncertainty principle is of the form $\Delta E \Delta t \geq h/4\pi$. When the energy of a system is measured, the product, $\Delta E$ times $\Delta t$, can never be zero, where $\Delta E$ is the minimum uncertainty in the measured energy, and $\Delta t$ the time needed for the measurement.

To measure the energy of a system, we have to interact with it. To obtain a definite value, some time is required for the interaction. The more time we have, the more precise the result can be. If only little time is allowed, the uncertainty increases.

This version of the uncertainty principle is demonstrated, for example, when the energies of excited electronic states in atoms or molecules are measured. When an atom absorbs a photon it uses its energy to make a transition to an excited (high energy) state. Such states are not stable but, after a short time, spontaneously drop back to the ground state. A photon is emitted in this process whose frequency corresponds exactly to the energy difference between the ground state and the excited state involved. If the lifetime of a state is very short, the time available for observing it is necessarily small ($\Delta t$ is small) and its energy cannot be measured precisely. Thus, when atoms in short-lived states return to the ground state, the energies of the emitted photons are not all exactly the same, but scatter about an average value. There is, as we say, *frequency broadening.* The more short-lived the excited state, the greater the frequency broadening of the transition. In general, if an event occurs within a very short time interval, its energy is uncertain. Time and energy of an elementary phenomenon cannot both be known exactly.

$E = mc^2$ tells us that energy and mass are equivalent. Therefore, uncertainty in energy is equivalent to uncertainty in mass. When only a very short time is available to look at an object, the uncertainty in mass may be so large that it is impossible to ascertain whether the thing existed or not. For this reason **continuous observation** of anything is impossible. Continuous observation is uninterrupted observation. It implies that a thing can be observed at a given time, $t_o$, and an arbitrarily short time interval later, at $t_o + \Delta t$, including the limit $\Delta t = 0$. *The uncertainty principle forbids this.*

In part I of the book we discussed the nature of facts. Facts are usually defined as elements of knowledge that can be verified by observation. The uninterrupted identity of things is usually taken for such a fact. Due to Heisenberg's uncertainty the *permanence of objects* is not an observable fact, but merely an assumption taken for granted.

Interestingly, in contrast to the quantum world, the ability to accept the permanence of unseen objects is essential to our development in the ordinary world. Most adults have no problem to understand that objects are independent of their perceptions and continue to exist when out of sight. But a developing child needs nearly a full year to learn *object permanence.* Hide a toy behind a screen and look, how amazing, it is still there. This is the discovery of **Jean Piaget**, Swiss psychologist, whose studies of childhood development have become standard textbook material.

Scarr et al. (1986) explain that *"For the infant in sensorimotor states 1 and 2, out of sight is not only out of mind, it is out of existence. A baby at this age tracks a moving object with his eyes, but if it disappears behind a screen, he loses interest almost immediately. If a magician covers a rabbit with a hat, then removes the hat to reveal an empty table, the baby is not the least bit surprised."* The true magic is not that the rabbit has gone, but that it should still be there.

It would not be reasonable to assume that ordinary things—trees, houses, the objects of everyday life—lead an interrupted existence, or an interruptible one. In contrast, the *components* of ordinary things, the elementary particles, as a rule do just the opposite, continually giving up their identity in ceaseless annihilations and re-creations.

## At the Foundation of a Stable World, Transitory Constituents

One of the strict conservation laws ruling all physical processes is the law of conservation of energy. Energy cannot be destroyed into nothing or created from nothing. What goes in, must come out. This law is obeyed absolutely; there are no exceptions, though *temporary violations* are allowed by the uncertainty principle. The examples presented in the following section were taken from the book by **Ford** (1963), which is an excellent survey of this field.

Assume that mass particles appear in empty space out of nothing. Since mass and energy are equivalent, such a phenomenon would violate the law of the conservation of energy and should be impossible. But now consider that the particles formed in this illegitimate way will immediately vanish after they come into being. In that case the time available to observe them can be so short that one cannot prove their existence with certainty, because $\Delta E$ is so large. Such a phenomenon would not be a clear violation of the conservation of energy, because it could not be precisely established. Thus, it is not forbidden.

There is a rule in particle physics (see Ford, 1963) according to which *"everything that **can** happen without violating a conservation law, **does** happen."* Thus, it is now believed that particles do form spontaneously in empty space out of nothing—*pairs of particles,* actually, to obey additional conservation laws, such as the conservation of charge—but they disappear so quickly that their existence cannot be demonstrated. These pairs are called **virtual particle**s. Herodotus, 400 B.C. (quoted in *Van Nostrands Scientific Encyclopedia,* 1947, p. 720): *"If one is sufficiently lavish with time, everything possible happens."*

The uncertainty principle allows violations of the law of conservation of energy, if they last only a very short time. Virtual pairs exist, but not long enough for anyone to really *know* of their existence. Again we encounter the unexpected **power of information**—**the mind-like aspect**—in elementary physical phenomena: *since it cannot be **known** for sure that it happened, the uncertain phenomenon is allowed.*

In this way elementary particles are *"centers of continual activity"* (Ford, 1963) by which incessantly virtual particles are created. Protons, for example, will spontaneously lose their identity by changing into neutrons and virtual pions (members of another family of elementary particles). If a proton is alone, the neutron/virtual-pion pair must quickly recombine to keep the energy balance in order. Ford (1963): *"Because the interaction responsible for this activity is so strong, these processes occur repeatedly . . . and the proton is surrounded by its cloud of virtual pions, darting this way and that, but leashed by the uncertainty principle to remain within little more than $10^{-13}$ cm of the nucleon core."*

Since protons are indistinguishable, it may be meaningless to ask whether the proton formed by recombining a neutron/pion pair is the same as the one that first perished in creating the neutron and pion. Nevertheless, the best interpretation of these phenomena is that when the neutron and virtual pion were formed, the act was *"catastrophic"* as Ford puts it, meaning that the original proton ceased to exist and a new proton was formed later on.

When a proton is not alone but close to another, real, neutron, as in an atomic nucleus, the second neutron may capture the virtual pion, and turn itself into a proton, leaving the originally virtual neutron stranded in reality. In this way the elementary particles—the constituents of ordinary and enduring things—continually are involved in annihilations and re-creations, giving up their identity, doing it with anyone, now ceasing to exist, and now coming into being.

In general, all interactions between particles are now believed to involve the creation and annihilation of matter. Lasting individuality of the components of permanent things is not the rule, but the opposite is. In instantaneous and localized creations, free particles even interact with themselves in continuous chaotic activities. Ford (1963):

> The strong hint emerging from studies of elementary particles is that the only inhibition imposed upon the chaotic flux of events in the world of the very small is that imposed by the conservation laws. Everything that *can* happen without violating a conservation law, *does* happen. . . . Briefly stated, the new view is a view of chaos beneath order—or what is the same thing, of order imposed upon

a deeper and more fundamental chaos. This is in startling contrast to the view developed and solidified in the three centuries from Kepler to Einstein, a view of order beneath chaos. . . . In this century the theories of relativity and quantum mechanics and the experimental evidence from the world of elementary particles have combined to reveal a deeper-lying and more fundamental chaos. Particles are found to have transitory existence; empty space is a bee-hive of disordered activity; an isolated particle is engaged in a constant frenzied dance whose steps are random and unpredictable; . . . The emerging picture of the world is that of a nearly limitless chaos governed only by a set of constraining laws, a world in which apparently everything that *can* happen, subject to the straitening effect of these conservation laws, *does* happen.

Apart from creating virtual pions, protons may be involved in whole networks of virtual pair productions, and electrons constantly create virtual photons. When a transition is made in an atom from a higher electronic state to a lower one, the energy difference can be transferred to one of the virtual photons which then becomes a real one. An electron hovering about the nucleus of an associated hydrogen atom may collide with an electron/positron virtual pair that formed out of nothing and annihilate with its positron. The energy released can be transferred to the leftover electron from the virtual pair, which then becomes the atom's new real electron. Individuality, if it has any meaning at all among indistinguishable particles, is constantly given up at the elementary level, one thing exchanged for another. When elementary cats vanish behind elementary sofas on one side, they vanish from the universe, and the cats that emerge on the other side are new ones—look alikes—different selves which replace the former ones.

One is also reminded here of Parmenides and his thesis that there is no vacuum. Due to the formation of masses in empty space, the vacuum constantly fluctuates. *"What is not cannot be. There is no vacuum."*

*Ordinary things have a definite identity. The indistinguishable elementary particles which make them do not. Ordinary things can remain in existence forever. The elementary particles which make them cannot. Ordinary things partake of a stable physical order. The elementary particles on which that order is based are involved in ceaseless, chaotic activities.*

*At the foundation of reality continued existence and identity are not the rule, but the opposite is, transitoriness and fleeting appearance in reality.*

# THE SURREALISM OF SUPERPOSITIONS OF STATES: THE CASE OF SCHRÖDINGER'S CAT

The surrealistic nature of superpositions of states is revealed when a system is suspended in a superposition of contradictory states. For example, the solution of Schrödinger's equation for a free particle can be expressed as a superposition of states in which the particle is moving along an axis in two ways, both to the right $\psi(\rightarrow)$, and to the left, $\psi(\leftarrow)$. In this composite state, momentum has no definite value. If you speak out of both sides of your mouth, you say nothing. If you move in opposite directions at the same time, you go nowhere. Thus, there is no motion and such a state could be adequately termed a **Parmenidian state**. In the same way, the traditional notions of motion lose their meaning, when an unobserved particle is nowhere because it can be everywhere.

If you are both here and there, you are nowhere. If you take all possible paths at the same time, you take no definite path at any time. In a superposition of states, no definite properties. Thus, when the concept is applied to ordinary things, paradoxical situations arise. To illustrate this point, Schrödinger invented what is now called **the Paradox of Schrödinger's Cat.**

In Schrödinger's paradox a live cat is locked up in a box with a bottle containing a deadly poison and an atom of a radioactive element. When the atom decays, the bottle breaks, the poison is released, and the cat dies. As long as the atom does not decay, the cat lives.

The decay process of a radioactive element is a random event—a quantum process—that is characterized by its half-life. After the half-life of an atom, the probability is ½ that it has decayed, and ½ that it has not yet decayed. Thus, after a time corresponding to the half-life of the radioactive element in the box with the cat, the probability is ½ that the cat is dead, and ½ that it is alive. Unobserved, the quantum cat is now in a superposition of states in which it is both dead and alive. As Schrödinger put it, *"The half-live and the half-dead cat are smeared throughout the whole box."*

According to some interpretations of quantum mechanics the wave function collapses and an actual situation arises, when an observation is made. But what does that mean—making an observation? If the measuring instrument, too, is considered a quantum system, its wave function will itself evolve in a superposition of states in which different outcomes of a measurement are possible. To relieve it from superposition, another instrument is needed, and for that one, yet another, and so on. There is an infinite chain of measuring devices, and the question is where $\psi$ will collapse.

In a variation of Schrödinger's cat experiment, the cat is replaced by a human observer, usually called **Wigner's friend**. Open the box and ask the friend what happened. The question is, in what state is the friend immediately before his mind is being probed? In a superposition of states? And what state will *you* be in, probing Wigner's friend, prior to being probed yourself by a friend of your own? At what link in the chain does the wave function collapse?

In a general way, the most reasonable assumption seems to be that the collapse occurs when a system in a superposition of states interacts with one in a state of ordinary reality. For the half-live and half-dead cat, the question as to whether or not it existed in a superposition of states before it was observed can be answered by a simple experiment: The first observer of the phenomenon should open the box and take a deep breath: *the stinking cat has been real dead for a long time.*

# Appendix 13

# THE PAULI PRINCIPLE

The origins of the Pauli principle are in the theory of relativity. The principle is an expression of the fact that the wave functions of atoms or molecules which contain several electrons change sign when the labels of two electrons are exchanged. Apart from the rigorous mathematical treatment, the meaning of this rather abstract principle can be illustrated in the following descriptive way.

Consider an atom that contains a number of electrons, say $n$ of them. To accomodate these electrons the atom must have a place, or a quantum state, for each of them. Thus, for $n$ electrons there must be at least $n$ individual places or single-electron states, $\psi_1$, $\psi_2$, $\psi_3$, ... $\psi_n$, symbolizing wave states as usual. The total wave function—or many-electron wave function—for the atom with $n$ electrons will then be some combination, for example some product, of the individual states, and we can write symbolically $\psi_{1,2,3...n} = \psi_1 \times \psi_2 \times \psi_3 \times ... \psi_n$, where "x" is the multiplication sign. In order to build such an atom from scratch, we have to take the appropriate nucleus and then fill one electron after another into the available single-electron states, putting the first electron into $\psi_1$ and denoting the result as $\psi_1(\text{I})$, putting the second electron into $\psi_2$, denoting the result as $\psi_2(2)$, the third into $\psi_3$, forming $\psi_3(3)$, and so on, up to $\psi_n(n)$. A hotel manager might assign his guests to rooms in a similar way; guest no. I to room no. I, guest no. 2 to room no. 2, and so on. For the atom under consideration the total wave function, then, resulting from the procedure is

$$\psi_{1,2,3...n} = \psi_1(\text{I}) \times \psi_2(2) \times \psi_3(3) \times ... \psi_n(n).$$

In performing this operation, we have assigned labels to the electrons, labeling one of them as electron no. I and assigning it to the first state; labeling another one as no. 2 and assigning it to the second state; no. 3 to the third state; and so on. Now, electrons are indistinguishable particles and the order in which they are picked and then labeled is entirely arbitrary. Instead of calling the first pick no. I, we could have called it no. 2, at the same time calling the second electron no. 1. Compared to $\psi_{1,2,3\ldots n}$ the wave function that results when the labels are exchanged on electrons 1 and 2 can be symbolized by

$$\psi_{2,1,3\ldots n} = \psi_1(2) \times \psi_2(1) \times \psi_3(3) \times \ldots \psi_n(n).$$

Since the operation of labeling indistinguishable electrons is arbitrary, *changing the labels on the electrons should have no observable effect.* In order to appreciate the consequences of this condition, remember that the wave function itself is not observable, but only the squared magnitude $|\psi|^2$; namely, $|\psi|^2$ represents probabilities for possible events to occur. Thus, if the exchange of labels on electrons in many-electron wave functions should not affect the square, it follows that the wave functions describing atoms must be of such a type that either $\psi_{1,2,3\ldots n} = + \psi_{2,1,3\ldots n}$, or $\psi_{1,2,3\ldots n} = -\psi_{2,1,3\ldots n}$. In the first case (plus sign) we say the wave function is *symmetric with respect to exchanging the labels on two electrons*; in the latter (minus sign), *it is antisymmetric.*

The Pauli principle says that the wave functions of real electrons and protons—spin-½ particles in general—are antisymmetric. *When the labels of any two identical electrons in the same atom or molecule are exchanged, the total wave function changes sign.* The principle is equivalent to stating that *no two electrons in the same atom can be in the same quantum state,* meaning that no two electrons in the same atom can be in states which have the same set of four quantum numbers (one for energy, one for the magnitude of angular momentum, one for the orientation of angular momentum, and one for the orientation of spin). Because of the last formulation Pauli's principle is often called the *exclusion principle.*

The principle sounds rather esoteric and academic, but it is in fact the basis for the existence of the Periodic Table of the elements, for the way molecules form by the bonding of atoms, for the way molecules interact to form gases, liquids, or solids, or the molecular aggregates active in living organism—in short *the Pauli principle is the basis for the observable order of the universe, including the existence of life.*

Margenau has given an excellent description of this principle (Margenau, 1984): *"The laws of symmetry open up vistas of an enormous scope, for every kind of order in the molar universe seems to be a consequence of them . . . As has already been indicated, Pauli's exclusion principle accounts for atomic structure. Without this structure, every atom would collapse into a positive nucleus surrounded by an unorganized mass of negative charges . . . Pauli's principle allowed the Periodic Table to be understood in one fell swoop. Associated with this understanding was the discovery of valence: the forces that unite atoms into molecules . . . Not only valence forces but even the weaker intermolecular forces that cause adhesion of substances, capillarity, and surface tension have their origin in the antisymmetry principle coupled with the symmetry-controlled forces arising from the arrangement of the outer electrons . . ."*

It is an important aspect of Pauli's principle (Margenau, 1984) that it is not enforced by any ordinary physical property that we know. When an electron enters an ion, somehow it *knows* the quantum numbers of the electrons which are there, and somehow it *knows* which atomic orbitals it may enter, and which not. The avoidance of occupied orbitals is not due to the electrostatic repulsion between electrons or some other mechanical property—it is just due to the antisymmetry requirement of $\psi$, another manifestation of the mind-like aspects of physical reality, similar to the power of information to affect observable physical states.

Different orbitals—single-electron states—coexist in the same part of space and are readily occupied by different electrons. Similarly, when two closed shell atoms approach that cannot form a chemical bond, a significant part of the aversion that develops is due to the fact that the electrons in one of the atoms somehow realize that the available quantum states in the other are already occupied and must be avoided. Naively speaking: the electrons are nearly extensionless dots while orbitals are spread out to infinity. That an arbitrarily large number of dimensionless dots are not allowed to populate the same orbital, at different ends of it as it were, does not make sense.

When two molecules or objects, A and B, are far apart, it is reasonable to say that the two sets of wave functions that represent them, $\psi_A$ and $\psi_B$, are separate and independent. In the simple molecule, $H_2^+$, for example, in which one electron binds two hydrogen nuclei (protons) together, the system can be described by two atomic wave functions, $\psi_A$ and $\psi_B$, when the nuclei are far apart. In contrast, when two objects approach to close proximity, the concept of separate states loses its meaning. There is strong overlap between $\psi_A$ and $\psi_B$, the states on A get mixed up with the states on B, they lose their identity and by their combination bonding (stabilizing) and antibonding (destabilizing) states are formed. It is at this point that the electrons recognize that some of the states are occupied and that they must retreat to destabilizing states, when the others are full. This is *orbital occupancy avoidance pressure*. It happens in general when two objects approach too closely so that their states begin to overlap and mix. Even though the atomic nuclei in molecules are essentially surrounded by nothing but empty space, they cannot be squeezed together at will; the symmetry of the quantum waves forbids this.

Like probability fields and elements of information, symmetry is a mental property. Knock your head against a wall and feel the mind-like nature of the foundation of reality.

Margenau (1984): *"I wish once more to impress on the reader (1) the vast range of application and the immense fruitfulness of a law regulating the observable called symmetry and (2) its abstractness, its lack of reference to any of the ordinary properties of matter. The distinguished biophysicist Harold Morowitz calls symmetry noetic, a term derived from the Greek work nous, which means mind or consciousness.*

*"There is indeed something quasi-mental, nonphysical, about it. Earlier I used the phrase 'one electron knows what the others are doing.' The amazing fact is that we do not know of any physical influence that effects the avoidance by one electron of an already occupied atomic state. Furthermore there is no evidence whatsoever that the adjustment to the principle requires time. We may have here a true case of action at a distance . . ."*

# Appendix 14

# THE EPR PARADOX

In their 1935 paper "Can Quantum-Mechanical Description of Physical Reality Be Considered Complete?" Einstein, Podolsky, and Rosen (EPR) pointed out that, in judging the success of a theory, we have to ask two questions: (1) Is the theory correct? and (2) Is the theory complete?

EPR define **completeness** as meaning that *"every element of the physical reality must have a counterpart in the physical theory."* Something is an **element of reality** *"if, without disturbing a system, we can predict with certainty the value of a physical quantity, then there exists an element of the physical reality corresponding to this physical quantity."*

Now, according to Heisenberg's uncertainty principle, *when the momentum of a particle is known exactly, the position has no physical reality.* Applying the definitions given above, a thought experiment can be designed to test whether the quantum mechanical view of Heisenberg's uncertainty is correct, or the quantum mechanical description of reality incomplete. The exercise was prompted by Einstein's deep dissatisfaction with the strangeness of quantum theory. While he appreciated the mathematical consistency and the utility of this theory, he also thought that it should be replaced by something that allowed a more reasonable view of reality. Thus, when it was impossible to doubt that quantum mechanics was *correct,* it was possible to prove that it was deficient because it was *incomplete.*

To perform the EPR thought experiment, consider two particles: because of Heisenberg's uncertainty, the position and momentum of particle 1, $x_1$ and $p_1$, are incompatible observables. Similarly, for particle 2, $x_2$ and $p_2$ are incompatible observables. They cannot be known exactly at the same time. However, the *relative* positions of the two particles on an axis, say the x-axis, or the distance between them, given by the difference, $x_1 - x_2$, between their coordinates, and their *total* momentum (the sum $p_1 + p_2$) are not subject to Heisenberg's uncertainty. That is, it is in principle possible that $\Delta(x_1 - x_2)\,\Delta(p_1 + p_2) = 0$.

Assume that the two particles at one time interact and then are separated so that information on the distance between them and on their total momentum is not disturbed. Then at a certain time, t, an observer can measure $x_1$, and will know $x_2$ because the difference $x_1 - x_2$ is known; or he can **choose** to measure $p_1$, and then will know $p_2$ because the sum $p_1 + p_2$ is known. Since either $x_2$ or $p_2$ is exactly predictable in this way, both must correspond to an element of reality. EPR: *"Since quantum mechanics forbids this, it must be incomplete."*

Taken from a different point of view: At a certain time t, one can decide to measure either the position or the momentum of a distant particle, using a local particle as the measuring device. This means that immediately prior to the chosen measurement, the distant particle must actually *have owned* both a definite position and a definite momentum, or *something I do here, now, has an instantaneous effect a long distance away.* Einstein, Podolsky, and Rosen concluded that such a violation of the locality principle was unacceptable: *"No reasonable definition of reality could be expected to permit this."*

In 1952 Bohm formulated a version of the EPR paradox involving the spin components of two spin-½ particles (such as electrons or protons, with s = ½) in what is called a singlet state. When two electrons form a singlet state, their angular momenta are counter-aligned in such a way that they cancel and the total angular momentum is zero. In terms of quantum mechanics, the total angular momentum quantum number is zero.

The condition that the individual angular momenta of the two particles cancel in the singlet state implies that, when the spin components of the two particles are measured along the z-direction and when the orientation of one of them is found up, the orientation of the other one must be down.

**If s(z,1) = z+, s(z,2) = z-**

**If s(z,1) = z-, s(z,2) = z+**

According to EPR, s(z,2) is an element of reality. However, instead of measuring s(z,1), one could have measured s(x,1). This would have predicted s(x,2) exactly; that is, s(x,2) is also an element of reality. However, in quantum mechanics, s(x) and s(z) are subject to Heisenberg's uncertainty, and when one is known the other one is not real. Thus, again, quantum mechanics is incomplete.

Choosing an alternative view: By measuring s(z,1), particle 2 is put immediately into a definite $s_z$ state, which means that it is in a superposition of $s_x$ states. By measuring $s(x,1)$, particle 2 is immediately put into a definite $s_x$ state, which means that it is in a superposition of $s_z$ states. Thus, *measurement on particle 1 has an instantaneous effect on particle 2, a long distance away. Locality is violated.*

Non-locality: If the universe is strictly local, no effect, no influence can propagate faster than the speed of light. In a non-local universe, instantaneous effects may act over arbitrarily long distances. Something that happens here, now, has an instantaneous effect at the other end of the universe; and vice versa. Non-local effects are "random" to human intelligence, but who wants to say what meaning they might carry for a higher intelligence?

The EPR paradox originally was intended to prove that quantum mechanics is incomplete. Instead it inspired the formulation of Bell's inequality and an entirely new field of coherence-experiments which have revealed that quantum systems that once interacted stay connected over long distances.

# THE NON-LOCALITY OF THE UNIVERSE

By the concept of *locality* we understand the notion that physical processes are affected only by local influences, by agents which are present at the location where the action takes place, or in the immediate vicinity, so that they can act in an instant. More precisely, the concept is connected with the proposition of relativity that no signal can travel faster than the speed of light. Thus, any process occurring in the depths of the universe can affect my own physical state only after *due time*, namely after a signal has had the time to travel from there to here at a speed not greater than the speed of light. Locality means that nothing which happens *there, now,* can affect me *now, here,* if any influence or signal traveling at the speed of light cannot get here from there in no time. This is known as *Einstein separability*.

In a *non-local* universe something that happens now, in a manner of speaking at the other end of the universe, may have an instantaneous effect on me, distance being irrelevant. Vice versa, an experiment performed here may exert an instantaneous influence on a system somewhere else, a long distance away. This process admits of effective agents propagating faster than the speed of light, or *superluminally.* There is a general notion that such phenomena should not be able to occur. In a paper by Einstein, Podolsky, and Rosen (1935) that deals with a paradox now called the Einstein Podolsky and Rosen (EPR) paradox, the authors state about superluminal processes that *"no reasonable definition of reality could be expected to permit this."*

In contrast, *Bell's theorem* and the supporting experiments indicate that the opposite is true, that *the nature of the universe is non-local.* Specifically, Bell's theorem shows that experiments can be designed in which the results obtained by an observer at some location in space may depend on operations performed by a second observer at another location, a long distance away.

# Bell's Theorem

Elementary particles such as electrons or protons are characterized by the fact that they possess an intrinsic rotational energy as though they were tiny balls spinning about an axis, like the earth is spinning about an axis through its poles. The axis of rotation defines a direction in space which is subject to a strange quantization: every time when the orientation of the spin axis of an electron is probed in the laboratory, it is found at a fixed angle, $\alpha = 54.7°$, with respect to a reference axis *that is selected by an observer*. This means that the spin of electrons is allowed *two quantized alignments*, forming the angle $\alpha$ either with the positive direction of a given reference axis or in opposition to it. By convention, the reference direction in the laboratory is often termed the *z-axis*, and we say that the spin of an electron can have one of two *quantized z-components, $z^+$ or $z^-$*. This effect, termed *space quantization*, is a true quantum effect. It is as inexplicable as it is unexpected, and the different orientations of spin are distinct *quantum states*. Experimentally, spin orientations are easily determined with so-called *Stern-Gerlach analyzers*, special magnets whose fields define a direction in space along which any charged particle will align its rotation.

It is a peculiar aspect of this phenomenon that fixed orientations of elementary particle spins are inexorably enforced with respect to *any* z-direction which an observer selects *at random* in the laboratory. For example, if an analyzer has been aligned with the *horizontal* direction of the laboratory, an electron passing this device will align to either the right or the left. If the same electron is then made to interact with a second analyzer whose special axis has been aligned *vertically*, at once the particle must take the quantum leap and align its spin in one of the states that a vertical analyzer allows; namely, up or down.

Bell's theorem concerns pairs of particles, such as two protons (call them $p_1$ and $p_2$) which at one time interact with one another, forming a special state called a *singlet state*, and then separate and move in opposite directions toward observers in two different laboratories ($L_1$ and $L_2$), which are a long distance apart from one another. An analyzer in $L_1$ will then measure the spin orientation of the first proton, $p_1$, along an axis $A$, while that of $p_2$ will be measured in $L_2$ along an axis $B$. In a singlet state the spins of the two particles are *antiparallel*. This means that, if axis $A$ and axis $B$ have the same direction in space, then each time when the spin is found up, or $A^+$, for $p_1$, it will be down, or $B$-, for $p_2$, and vice versa. This result will be obtained no matter what direction is chosen, *at random*, for axis $A$ and axis $B$. As long as the two axes point in the same direction, invariably $A^+$ will be correlated with $B$-, and $A$- with $B^+$. *Perfectly negative spin correlation* is found for particle pairs in the singlet state.

An interesting situation arises when the spin components of $p_1$ and $p_2$ are measured along two axes $A$ and $B$ which point in *different* directions in space. In this case, the results obtained in $L_1$ ($\mathbf{A+}$ or $\mathbf{A-}$) and $L_2$ ($\mathbf{B+}$ or $\mathbf{B-}$) will still be correlated, but the correlation will be more complex than in the simple case when axis $A$ and axis $B$ point in the same direction; that is, the spin correlation varies with—it will be some function of—the angle between the two axes.

On the basis of some technical details, some of which are described in Appendix 16, such spin correlation functions can be derived for singlet pairs and experimentally tested. Without considering the details at this point, it is easy to see that the results of the measurements will depend in a fundamental way on whether the twins of a singlet pair become *totally independent* of one another when they move apart, or whether they stay in some way *connected*, no matter how far apart they are. In the first case—*pairs of protons with disconnected twins*—the observations in $L_2$ on $p_2$ will be independent of the operations performed in $L_1$ on $p_1$. In the second case—*protons separated in space but intimately connected*—the direction chosen for $A$ in $L_1$ and the outcome of the experiment, $A^+$ or $A$-, may *instantaneously* affect the state of the system in $L_2$. It is obvious that different correlation functions should be observed for the former than for the latter. Putting it crudely, if two protons are tied together by a long string and experiments on one will deliver a jerk on the other, the observed orientation correlation will differ significantly from that which will be obtained when no strings are attached.

This is exactly what Bell's theorem is about. Examining pairs of particles in the singlet state whose spin components are measured along different directions in space—say along three directions $\mathbf{A}$, $\mathbf{B}$, or $\mathbf{C}$, which are not necessarily mutually perpendicular like the axes of a Cartesian coordinate system—and assuming that *no* superluminal ties exist between remote quantum particles, Bell derived an expression for spin correlation, ***Bell's inequality,*** that predicts the relative frequencies of various possible combinations of spin orientations— $A^+$ or $A$-, $B^+$ or $B$-, and $C^+$ or $C$ -—found for twins in singlet pairs in series of repeated measurements.

Note this important condition of Bell's inequality: *it was derived on the basis of a local realistic view of the world (Espagnat, 1979). That is, (1) the particles were assumed to **own** their physical properties regardless of any observers; (2) the validity of induction was assumed; and (3) Einstein separability was assumed. If a different view had been adopted, an entirely different inequality would have resulted than the one given by Bell.*

It is a second important aspect of Bell's inequality that it can be tested by experiments. Thus, essentially, by comparing the predictions made by this theorem with the results of appropriate experiments, *the validity of the local realistic view of the world can be empirically tested.*

**When the experiments are performed, involving large numbers of pairs of particles in the singlet state, and the numbers of different outcomes are counted, some directions in space are found along which Bell's inequality is violated.** That is, for some orientations of the axes **A**, **B**, **C**, combinations of spin directions—up or down along **A**, or **B**, or **C**—are found with relative frequencies that are higher than Bell's theorem will allow. Thus, one is forced to conclude that at least one of the assumptions underlying the local realistic view of the world is in error. At present the most widely held view is that Einstein separability is violated, implying that the results of physical experiments performed at one location in space may depend on experimental conditions selected at random a long distance away.

## Non-Locality Derived from the Extended Nature of the Quantum Probability Waves

Stapp (1993) has shown that Bell non-locality is directly related to the extended character of quantum waves: when the paired particles leave the interaction center where the singlet state was formed, the wave function $\psi$ that represents this system will spread through the region of space that contains the laboratories $L_1$ and $L_2$. The potentialities of this case entail that $\psi$ is a superposition of two components. One of them corresponds to the state in which the spin of $p_1$ is up and that of $p_2$ down along a certain direction, while the second component represents the state in which $p_1$ is down and $p_2$ up along the same direction. As usual, $\psi$ will collapse and an actual state spring into being from the superposition of possible ones when an observation is made. Exactly how the wave function will collapse will depend on how the observer who makes the first measurement chooses the axis direction, and what result, up or down, he obtains. As soon as $\psi$ has collapsed for one of the particles, *in that same instant* the probability amplitude for the second one is changed accordingly, no matter how far away it is.

The conclusion is, as put by Stapp (1993): "Bell's theorem [demands] that what happens in one spacetime region must, in certain situations, depend on decisions made in spacelike-separated regions. . . . The conclusion is simply that there is no way for nature to select results that are compatible with both the predictions of quantum theory and the condition that the results observed in each region be independent of the choice of experiment made in the other region. . . . no deterministic theory can exclude faster-than-light influences, if it is to reproduce the predictions of quantum theory."

Connecting Bell-non-locality with the nature of the quantum waves implies that $\psi$ is understood, Stapp (1993), *"to represent real tendencies for responses of devices or observers."* This is in contrast to the view that *"the wave functions describe the evolution of the probabilities of the actual things, not the actual things themselves."* If the real-$\psi$ interpretation holds and a measurement *here* causes the instantaneous collapse of a **real** tendency *a long distance away,* this process involves *faster-than-light information transfer.* The instantaneous transmission of information over long distances, in this way, is not a violation of Einstein's relativity. Stapp (1993): *"What Einstein forbade was faster-than-light signals, where a signal means a controlled transfer of information. The same quantum-theoretic rules that lead to the apparent necessity of faster-than-light information transfer exclude the possibility of faster-than-light signals."* Bell-transmission of information is without control, because collapses of the wave function and creations of actual states are random events.

The discovery of the non-locality of the physical universe is one of the most important discoveries of the history of physics. *Quantum particles which at one time interact are intimately connected by influences which are not attenuated or delayed, like gravitational or electrostatic forces, by spatial separation.* The importance of this discovery for the relation between the physical sciences and religion is obvious. In the classical universe God was a strangely controversial figure at best: either a constant breaker of His own laws, or an effective agent not truly divine but bound by travel at limited speed—not faster than the speed of light. Both aspects contradict our intuition of Divinity. In particular, it seems offensive that God, tending to some business in some part of the universe, should need billions of years to reach other parts where the plumbing needs fixing. *In contrast, the quantum phenomena have made it possible to believe again, **within** the framework of the physical sciences, in the presence of all-pervading, instantaneous, and all-powerful agents.* Thus, whereas physical reality has turned out to be disturbingly different from what we always thought, it has also revealed itself as being more in accordance than we could have hoped for, with our most intimate, spiritual needs.

A quote of lost origin: *The universe is network, not clockwork.*

# SOME TECHNICAL DETAILS
# CONCERNING BELL'S THEOREM

## *Quantization of Angular Momenta*

### VECTORS

Among the dynamic physical attributes of things are those that are fully characterized by merely specifying an amount, or a magnitude, such as for example the energy or the mass of a system, and there are those that need both a magnitude and a direction to be fully defined, such as for example a force acting on a mass. In dealing with forces it is not only important to know how strong a given force is, but also the direction in which it acts. In physics, directed quantities that carry both a magnitude and a direction are called **vector** quantities.

### ANGULAR MOMENTUM

In a closed system of moving objects, one of the constants of motion is linear momentum, the product $mv$ of mass, $m$, and velocity, $v$. Since $mv$ is a constant of motion, we say that in every mechanical process *linear momentum is conserved*. Similarly, when a particle with mass $m$ is moving in a circle with radius $r$ and its velocity is $v$, its motion is characterized by a quantity called *angular momentum*. Angular momentum is also a vector quantity and *a constant of motion;* that is, in every mechanical process angular momentum is conserved. For the particle moving in a circle the magnitude of angular momentum is given by the product $mvr$, mass times velocity times radius, and its direction is along the axis of rotation. It is seen from this formula that, the faster a greater mass moves in a larger circle, the greater its orbital angular momentum. Similarly, the faster a bigger spherical mass spins about an axis—like the earth is spinning about an axis through its poles—the greater its spin angular momentum.

## QUANTIZATION OF ANGULAR MOMENTUM

It was one of the unexpected discoveries of this century that masses cannot revolve in an orbit or spin about an axis with arbitrary velocities; they can do so only at certain speeds. More precisely, **angular momentum is quantized**. In addition, common elementary particles, like electrons, protons, or neutrons, were unexpectedly found to carry an intrinsic spin angular momentum, as though they were spheres spinning about an axis like a top. This intrinsic momentum is also quantized, being allowed a single, fixed value for each type of particle. Furthermore, when its direction is probed with respect to an axis in the laboratory, **only a limited number of orientations with respect to that axis is found to exist**. *We say that orientation is quantized and often refer to it as space quantization.*

For a classical particle, the amount of spin can be measured in terms of speed, but for the point-like quantum particles speed of spin has no meaning and magnitude of spin can only be measured in terms of angular momentum. If we use the magnitude of the spin of photons as a unit, assigning it the value of 1, then most particles either have 0, $\frac{1}{2}$, or 1 unit of angular momentum. Electrons and protons are spin-$\frac{1}{2}$ particles. Invariance of spin means that the amount cannot be changed; that is, it cannot be accelerated nor stopped. **Spin is an intrinsic property that elementary particles own, regardless of observation.**

## SPACE QUANTIZATION

Space quantization is an amazing phenomenon. We can take an electron as a specific example. Each electron has intrinsic angular momentum with fixed magnitude $(\sqrt{3})(h/4\pi)$, where h is Planck's constant. In addition, when the components of this spin are measured along a direction in space, only one of two possible values can be found, either $+h/4\pi$ or $-h/4\pi$, showing that only two orientations are allowed for this vector property with respect to a given axis, one up and one down, respectively. Whenever an axis is chosen in the laboratory, no matter in what direction it is pointing, the spin angular momentum (SAM) vector of the electron must orient in space in such a way that its projections along this axis assumes one of the two allowed values, one pointing up, ↑, the other one down, ↓.

A question that immediately comes to mind concerns the meaning of the phrase *"given axis in the laboratory."* What is that given direction or axis, often denoted as the *z-axis* or *z-direction*, who gives it, and who decides which way it points? The answer is that *the free choice of the observer decides*. The observer selects a direction in space along which the orientation of an electron's SAM is probed. The electron then must orient in such a way that the z-component of its SAM is either $s_z = +h/4\pi$ or $s_z = -h/4\pi$. If, next, the observer changes his mind and chooses another direction, *that same instant* the electron must realign because it must now be either up or down in the quantized way with respect to the new direction. An important aspect of these phenomena is that *different components of SAM correspond to different quantum states which can be separated in space*.

Quantum states are characterized by quantum numbers. For readers not deterred by quantitative expressions we will note that the magnitude of the SAM vector, $|\mathbf{S}|$, is determined by the spin quantum number s, following the expression $|\mathbf{S}| = [s(s + 1)]^{1/2} h/2\pi$. Only the value $s = \frac{1}{2}$ is allowed to electrons, so that, in this case, $|\mathbf{S}| = (\sqrt{3})(h/4\pi)$. In addition, the components of an electron's SAM along say, the z-axis, $s_z$, are quantized and determined by a quantum number called the *orientation quantum number*, usually denoted by $m_s$. Two values are allowed, $m_s = +\frac{1}{2}$ and $m_s = -\frac{1}{2}$. The possible components of angular momentum along the z-axis are, for electrons, $s_z = m_s(h/2\pi)$; that is, $s_z = +h/4\pi$ and $s_z = -h/4\pi$, where h is Planck's constant. It is in this way that, along the z-axis, the electron's spin is either up, $s_z$ and $m_s$ positive; or down, $s_z$ and $m_s$ negative.


THE STERN-GERLACH EXPERIMENT

Space quantization was first demonstrated in 1921 by the historic *Stern Gerlach experiment*, in which Stern and Gerlach directed a beam of vaporized silver atoms through an inhomogenous magnetic field. In an evacuated container, silver metal was vaporized and an atomic beam formed by letting the gaseous atoms effuse through a series of small holes into a larger chamber in which they passed an inhomogeneous magnetic field established by the differently shaped shoes of a magnet, one pointed, one round.

All atoms contain electrons whose individual angular momenta contribute to the total atomic angular momentum vector. In adding vector quantities, not only the magnitudes but also the directions have to be taken into account. For example, if two forces with equal magnitudes act on the same particle in exactly opposite directions, then the resulting force (the sum of the individual ones) is zero. In the case of a silver atom, the angular momenta of 47 electrons add in such a way that the resulting vector of the atom is that of a single electron. As a consequence, the angular momentum z-component of a silver atom can assume only one of two possible values with respect to a given axis in the laboratory; one pointing up, ↑; one down, ↓. Electricity is correlated with magnetism. Moving electric charges engender magnetic fields. Spinning electrons engender magnetic fields, and silver atoms, carrying the spin of a single electron, are small elementary magnets. Thus, when a beam of silver atoms passes through the poles of a magnet, the atoms experience a force whose strength and direction depend on their orientation in the magnetic field. Essentially, the direction of the field of a magnet defines a z-direction in the laboratory.

According to classical physics, elementary magnets like silver atoms can assume any arbitrary orientation in an external magnetic field. Thus, classical atoms in a magnetic field will be pulled every which way, and a beam of such atoms will be broadened—getting out of focus. Real silver atoms are different. Since their angular momenta can align in only two ways along a given z-direction, these little magnets have only two allowed orientations in the magnetic field. Thus, they are pulled into just *two* opposite directions; that is, a beam of silver atoms in an inhomogeneous field splits into two. **This is the result that Stern and Gerlach obtained.**

## ANGULAR MOMENTUM AND HEISENBERG'S UNCERTAINTY

An ordinary vector is defined by components along three axes in space. These axes are usually taken to be perpendicular, such as the Cartesian x, y, and z axes. For vector quantities, such as linear momentum, force, or velocity, all three components have definite values at all times. *For spin vectors this is not possible, because their components along different axes are subject to Heisenberg's uncertainty principle.*

It is yet another form of Heisenberg's uncertainty principle that, if the component, $s_z$, of the SAM of a particle along the z-axis is known, then the perpendicular components, $s_x$ and $s_y$, are not knowable. If a system is in a definite $s_z$- state, it is in a superposition of $s_x$- states in which the two possibilities, $s_x = +h/4\pi$ and $s_x = -h/4\pi$, have an equal probability to occur. Similarly, if a particle is in a definite $s_z$-state, it is in a superposition of $s_y$-states.

Because of Heisenberg's uncertainty, no experiment can give simultaneously exact values for the three components of the spin angular momentum of a particle. If, for example, spin-$\frac{1}{2}$ particles have passed a z-analyzer in the up channel, they are in the definite $z^+$-state and all will pass a series of additional analyzers in the $z^+$-channel. However, when the particles in the $z^+$-state subsequently are sent through an x-analyzer, one half of them will emerge $x^+$, the other half, $x^-$. If one now regards the particles in the definite $x^+$-states, one must not assume that they are also still in the $z^+$-state. Indeed, when they are sent again through a z-analyzer, they will come out at random through $z^+$ and $z^-$, because the $x^+$-state corresponds to a superposition of $z^+$- and z-states. *When a particle passes a Stern-Gerlach analyzer, the spin orientation state is altered.*

## Additional Data on Bell's Theorem

J. S. Bell showed that, if the universe is *local-realistic*, EPR-type experiments must yield results which are in contrast to quantum mechanics. Since the inequality that he derived can be tested experimentally, it is possible to test whether the universe is local and quantum mechanics in error, or quantum mechanics is true and the universe *non-local*.

A **local-realistic theory** of the universe is based on three assumptions defined by Espagnat (1979) and Polkinghorne (1985) in the following way:

(1) **Reality Assumption**: *At least some properties of the world have an existence independent of human observers.* That is, the regularity in physical phenomena is due to an underlying reality independent of observation.

(2) **Validity of Induction**: *Inductive inference is a valid mode of reasoning and can be applied freely.* That is, it is possible to predicate conclusions about *all* the members of a class, if a consistent set of observations exists on *some* of them.

(3) **Locality Assumption**: *No influence from a measurement made with one instrument on a measurement made with another can propagate faster than the speed of light.* Locality is also often called *Einstein locality* or *Einstein separability.* If influences can travel faster than the speed of light and an event at A instantly affects an event at B an arbitrary distance away, separability is violated.

To derive Bell's inequality, pairs of particles are considered, such as two protons, $p_1$ and $p_2$, which are in a singlet state. In singlet states, the spin angular momenta of the particles are aligned in such a way that the total spin angular momentum is zero. That is, the total spin angular momentum quantum number of the system, $p_1$ and $p_2$, is $S = 0$. Thus, when the components of the angular momenta are measured along a single direction, say along the **A**-axis, there is completely negative spin correlation. If $A^+$ is found for $p_1$, $A^-$ will be found for $p_2$, and vice versa.

In Bell's thought experiment many pairs of particles, $p_1$ and $p_2$, are prepared in the singlet state. For one pair after another, the individual twins are then separated and made to travel in opposite directions to two analyzers which are so far apart from one another that a measurement made by one cannot affect the other by any influence that travels at no more than the speed of light. In their respective laboratories the two analyzers can measure the spin components of $p_1$ and $p_2$ along *three different directions* in space, call them the **A**, **B**, and **C** axes. Thus, for each of the twins one of six spin components are possible, which point either up ($A^+$, $B^+$, or $C^+$) or down ($A^-$, $B^-$, or $C^-$) along the three axes. When measurements are performed on a large number of pairs, in each pair the spin orientation of one particle is found opposite to the other in every instant in which both are probed along the *same* direction in space—no matter whether that is **A**, **B**, or **C**. Alternatively, if the spin orientations of the twins are measured along different directions, we denote by $n(A^+, B^+)$ the number of times that, in a series of measurements, $A^+$ was found for one, while $B^+$ was found for the other. In the same way we define the numbers $n(A^+, C^+)$ and $n(B^+, C^+)$.

Now, Bell has shown that, *if one adopts the local realistic view of the universe*, then by simply applying the rules of combinatorial mathematics one can derive an inequality regarding the relative magnitudes of these numbers; that is, one finds that $n(A^+, B^+) \leq n(A^+, C^+) + n(B^+, C^+)$, which is a form of Bell's inequality. The formula shows that *Bell's inequality is an expression for the relative frequency of occurrence of combinations of spin components obtained in series of measurements along three different directions,* **A**, **B**, **C**, *for particle pairs in the singlet state.* The version of Bell's thought experiment presented above was taken from Polkinghorne (1985), who gives a detailed derivation of the inequality in this form; for equivalent presentations, see Espagnat (1979) and Rae (1986). *The formula holds, provided that **(1)** the spin components are elements of reality—that is, they are owned by each particle no matter whether anyone is looking at it or not; **(2)** the principle of induction is valid; and **(3)** Einstein separability holds.*

The experiments testing Bell's inequality typically consist of generating large numbers of particle pairs in singlet states, emitting individual twins to their respective detectors, selecting three directions for measurement, and adding up the numbers of interest after a series of such measurements. *When the experiments are performed and the numbers counted, directions in space can be found for which Bell's inequality is violated. That is, for some orientations of A, B, and C, some combinations of spin components which should occur less often than others actually occur more frequently.*

The violations of Bell's inequality show that some argument in its derivation is not in agreement with the nature of reality. Which one will it be? Is it the Reality assumption, Validity of Induction assumption, or Einstein Separability assumption, or are several of them at fault? The experiments are silent as to who the culprit is. The most widely accepted view at present is that the locality assumption is at fault, implying that **the nature of the universe is non-local**. The universe is made up of quantum systems whose coherence is not affected by their separation in space.

# Appendix 17

# THE EMERGENCE OF HISTORICAL PHILOSOPHICAL VIEWS IN QUANTUM ONTOLOGY

## I. *Anticipated Principles*

Ideas related to concepts of empirical physical science were often conceived a long time before the supporting evidence was known. This is the phenomenon of *anticipated principles.*

Anticipated principles are defined as concepts that relate correctly to a space-time structure of physical reality, but they do so in a way—frequently outside of the context originally assumed—which could not have been foreseen at the time of their conception, when the relevant observations did not yet exist.

The *concept of the atom* was invented by Leucippus and Democritus about 400 B.C., long before it was empirically established by modern physics and chemistry. In the sixth century B.C. Anaxagoras declared that the Sun was nothing but "a glowing stone" (and was promptly tried for blasphemy) anticipating the *homogeneity of matter* throughout the universe long before any evidence for it had been found. The Pythagorean theory of *fundamental numerical relations*, rather than substances, as the foundation of order in the universe, was developed a long time before quantum phenomena provided supporting evidence. Mystics of all times thought that *all is one,* in agreement with the connectivity of the quantum world. The properties of the imaginary wave functions of quantum mechanics are difficult to describe in words, reminding us of Lao Tzu's teaching that *eternal truth cannot be expressed in words.* Modern genetics is in agreement with both Parmenides *(there is no change)* and Heraclitus *(everything is in flux)*: the process of copying a gene is absolutely invariant and, hence, there is no change or becoming; and yet, errors in the molecular transcription of genes occur, leading to variations in species and evolution.

The ability to anticipate empirical principles prior to the discovery of the supporting facts is an amazing power of the human mind. It is as though for any basic postulate that an inspired mind expressed, later research will discover the appropriate context, not necessarily known at the moment of conception, to which it applies. Thus, one is tempted to modify the principle of quantum mechanics according to which *a quantum process that is not strictly forbidden must occur* and apply it to mental processes: **a mental process that is not strictly forbidden must correspond to an element of reality**.

It is a consequence of the phenomenon that, in studying the grand philosophical systems of history, it is often not so important to debate whether or not a valid basis exists for anticipated principles, but rather, whether or not a context can be found in which the essence of a thesis, not usually the details, corresponds correctly to an element of reality. In this sense philosophical systems may contain information that is meaningful, even though it cannot be ascertained, but errors arise when the details are taken verbatim. Since the empirical support is missing, anticipated concepts usually appear with a growth of unrealistic details. The reasons for Aristotle to deny the existence of a vacuum *(there is no proper place in it)*, or for Parmenides *(what is not cannot be)*, are not acceptable. Nevertheless, the thesis itself—*there is no vacuum*—has been confirmed by the quantum phenomena. The alchemists' rule that *the shape of a vessel determines the outcome of a chemical reaction* seems totally unjustified. Nevertheless, for a **particle in a box** whose size is comparable to molecular dimensions, the quantum mechanical analysis yields energy states and space distributions—properties affecting its reactivitiy—that depend on the shape of the box. Historical errors typically occurred when enough care was not taken to separate the essence of anticipated principles from the irrelevant details.

The principle, *what is reasonably anticipated must exist*, seems to be refuted immediately by the fact that many **contradictory** postulates were often anticipated in a reasonable way. However, the example of Parmenides and Heraclitus, the establishment of both chance and necessity, or of evolution and invariance, as valid principles of nature demonstrates that contradictory concepts can both contain verifiable components.

Anticipated concepts are a particularly striking element in the realm of quantum mechanics, where interpretations of quantum facts have revived a large number of seemingly strange concepts which the Grand Philosophical Systems of our history expressed a long time ago. We recall:

- Quantum mechanics suggests that *the world of elementary particles is different in essence than ordinary reality.* Anaximander (6th century B.C.): All things come from a primal substance, *but that substance—the apeiron—is not like any substance we know.* Like the Heisenberg state of the universe, the apeiron is infinite, *"encompasses all the worlds,"* and constantly creates new realities.

- In quantum mechanics, the *electrons in atoms are standing waves,* probability amplitudes whose interference creates the visible order of the universe. Plato: *the atoms of the elements are not things but mathematical forms.* Thus, the wave functions of quantum mechanics are examples of Plato's ideas. They are different from ordinary reality, but contain all of reality.

$\psi$ = *eidos* (image), the common essence of different things of the same kind. Real things only represent the truly real proto-images (ideas). The ideas are only imaged, made observable, by real material things.

- In a Heisenberg event, the transition from the *"possible"* to the *"actual"* takes place when an observation is made. Thus, *reality is created by observation.* Bishop Berkeley: "To be is to be perceived . . . Real things are ideas imprinted on the senses by the author of Nature."

- According to Heisenberg's ontology, unmeasured reality is one of possibilities or tendencies to come into existence. Aristotle: *Unformed matter is not truly real, but has the potential—potentia—to become real. Form gives matter reality.* Quantum mechanics: *observations give tendencies reality.*

Aristotle called *"hyle"* the stuff that *"had the potential to be."* Cicero later translated *"hyle"* into the Latin "materia."

- Unobserved $\psi$ contains all possibilities and is not an image of the reality that we know. Kant: *We do not have experience of things but of the appearance of things. What the "things in themselves" are, we can never know.* We cannot know $\psi$ but only the forms of it that our interactions create.

- According to an important principle of quantum mechanics, nothing is known about a system when it is not being observed. *No statements should be made about what happens in a system in between observations.* Do not think that the electron went through a particular slit, or went anywhere at all, if no observation was made. Do not think that a free particle owns a position in space, do not speculate where in the world it might be when the conjugate variable, momentum, is known exactly. **Positivism**, **Pragmatism**: *Stick to the positive, the observable facts, do not speculate on the nature of things.*

- According to quantum mechanics, the nature of reality is characterized by quantum coherence. Non-local instantaneous quantum effects, like those due to Bell non-locality, are essential features of this reality. David Bohm (1980): "*. . . reality is one unbroken whole, including the entire universe with all its 'fields' and 'particles.'*" Parmenides: *All is one.*

The ability to anticipate empirical concepts before they are quite real tempts one to suggest that the **human mind is amenable to inspirations from an invisible source**, to secret non-local influences, like quantum systems are sensitive to Bell non-locality. The human mind is like an antenna, as it were, for the reception of anonymous instructions that guide the evolution of its thoughts.

Historic examples of sudden insight and inspiration are many: During one night, **Descartes** saw in a flash the order of the universe. **Stevenson** looked for years for a story of man's double being. The story of Dr. Jekyll and Mr. Hyde was revealed to him in a dream. **Kekulé**, in his sleep dreamed of the structure of benzene for which he had searched for a long time. **Rilke**, after decades of painful inability to be productive as a poet, within a few days and nights was rushed by the Duino Elegies.

By formulating the principles of heredity, **Gregor Mendel** (1822–1884) founded modern genetics. It was Mendel's idea that, in genetics, the ruling principle is *all or nothing*. It has often been pointed out that Mendel's data were statistically not sufficient to support his conclusions and his experiments were designed by someone who knew the answer before the results; that is, an **inspired mind**.

In addition to the parallels summarized above, many other links can be found between the worldview of quantum mechanics and historical systems of philosophy. It is the purpose of this appendix to describe some of them. Like few other events in the history of mankind, the discovery of the quantum world has been a source of deep insight into the nature of reality, and many ancient ideas have found a new meaning and unexpected significance.

## II. Zeno's Paradoxes and the Uncertainty Principle

PROBLEM:

How can the continuity of motion be related to a sequence of stationary points in time and space, to discontinuous instants (like a movie that is composed of a sequence of still pictures)? How is motion possible as a continuous stream of separate motionless instants? Where are the movie actors between two frames?

ZENO OF ELEA (490–430 B.C.):

How can movement be understood if an object in motion at a specific moment *occupies* a specific place in space at which it is at rest at that moment? *We generally assume that a given object is somewhere. When we say it is somewhere, we mean it is at rest there.* If it is at rest at each point of its trajectory, and at rest at each point in time, the thing must be at rest all the time. That is, there is no motion. Vice versa, if in motion, the object must leave its place at a given time to reach the next place at a later time. *Where is it in between two places?*

One of Zeno's paradoxes is the flying arrow. *"The flying arrow cannot reach its goal."*

To reach G, its goal, the arrow must first reach H1, the halfway point, between G and S, the starting point. (H1 is the first halfway point.) Once at H1, to proceed to G, the arrow must first reach the second halfway point, H2, of the distance between G and H1. Between H2 and G is another halfway point, H3, which must be passed first to arrive at G. Between H3 and G is H4, then H5, H6, and so on. *There is an infinite number of halfway points between the position of the arrow at a given time and G.* No matter where the arrow is, at every given instant the distance to G has a halfway point which must be passed before G is reached.

Thus, Zeno concluded, the flying arrow cannot reach its goal. No object can move through an infinite number of points in a finite amount of time. Each half of an extended stretch of space is again an extended stretch of space, admittedly smaller, but of some extension.

Aristotle called Zeno's paradoxes *"aporiai"* (Russell, 1978). *"Zeno described four aporiai concerning motion, which caused a great deal of difficulty to those who like to explain them: the first is that there can be no motion because what moves first must reach the halfway point of its path before it can reach its end."*

"Aporia" (singular) in Greek means "pathlessness"; "lack of being able to go from one place to another"; a logical difficulty which there is no way out of; a perplexity, a confusing situation. "Aporos" means impassable (French "impasse"; a tie, no side wins). An aporia arises when arguments can be put forward both in support of and in opposition to a given proposition. "Aporetics" is the systematic discussion, or analysis, or science, of aporiai.

**Aporia**: In order to use the concept of a *continuous line* in mathematics, a line must be divisible into arbitrarily small segments. This is the essence of continuity. It implies that any given point can be defined with any wished accuracy (any number of significant figures). Infinite division of the segments of the straight line ultimately leads to dimensionless (non-extended) points.

Contrast: *the points are discrete, non-continuous.* The line is continuous and extended. How can discrete points form a continuous line? How can non-extended points, no matter how many, make an extended and coherent stretch of space? This is the *contrast between the discrete and the continuous. As between wave and particle.*

### The Quantum Zeno Effect

Zeno's first paradox has found an interesting equivalent in quantum physics—called the *quantum Zeno effect* by Sudarshan and Misra in 1977—that involves the **observation-induced paralysis of an atom when it makes a transition from one energy state to another**.

The quantum Zeno effect is the phenomenon that irradiated atoms will be prevented from making a transition from the ground state to an excited state if they are being watched at short intervals while the transition is going on. Experiments of this kind were performed by Itano and his group in 1989 (see the summary by Powell, 1990).

When an atom in its ground state is irradiated with photons that it can absorb, it has a high probability to use the photon energy to undergo a transition from the ground state to a higher energy, excited state. During such a transition from one energy level to another the wave function of the atom must evolve into a superposition of states that contains both the initial and the final state. Out of this superposition the atom can take the quantum jump and emerge in the excited state. Thus, when an atom in this situation is constantly being observed—that is, measurements of its energy are performed at a high frequency (e.g., by using short laser pulses)—the superposition has no time to evolve as needed for the transition, and the transition does not occur. The atom is paralyzed, as it were, by the frequent probing, which destroys the coherence of the ground state plus photon and excited state, and the wave function constantly collapses back to the initial state.

Another aspect of the flying arrow—called Zeno's third paradox —is its **anticipation of Heisenberg's uncertainty principle.** According to the argument, the arrow cannot fly because at every instant it occupies a specific place; that is, it *is* at a definite point in space. To **be** at a place means to be **at rest** at that place. *Occupying a point means to be at rest at that point.* To be somewhere means to be at rest there. Not occupying a point means to be nowhere. *Since the arrow is somewhere all the time, it is at rest all the time. Thus, there is no motion.*

**Aporia: There is motion. Versus: There can be no motion.**

Zeno's third paradox is in conflict with the uncertainty principle because it involves observations of a moving object at precisely determined points along a precisely defined trajectory. In reality, it is not possible to divide a trajectory into arbitrarily small moments of space and time. At a very limited point in space $\Delta x = 0$, and nothing is known about the velocity (the momentum). Vice versa, the projectile does not have a position; that is, it is nowhere, when its momentum at a specific moment is known precisely. Thus Zeno's third paradox asks for the impossible and the quantum phenomena reveal its untenable basis.

If the line of flight is divided into smaller and smaller segments, $\Delta x = 0$. This reduction of segments also demands an increasingly accurate determination of momentum. Thus, also $\Delta p = 0$. Vice versa, if it *is* at a definite point, $\Delta x = 0$; if it is at rest there, then $\Delta p = 0$. All this is not allowed. **The logical difficulty arises because a physical principle is violated** (Pietschmann, 1980). Zeno's third paradox may be a problem of logic, but not of the physical reality.

Zeno's third paradox also violates the second principle of logic, the principle of contradiction. *(A moving object is both at rest, because it is somewhere; and in motion.)* Hegel (see Pietschmann, 1980, p. 28): *"Something is alive only insofar as it contains a contradiction in it."*

## III. Nominalism/Realism and the Meaning of ψ

Paradoxes like *Schrödinger's cat* or *Wigner's friend* illustrate the need to discuss the question of *what the reality is of* ψ? ψ is a general wave form, the general concept of a system, its universal form. *Does ψ possess physical reality or does it exist only in our minds?* Do the wave functions represent physical systems, or only our knowledge of such systems? This is exactly the kind of issue that was at the roots of the conflict between Medieval Realism and Nominalism. That is, **what kind of reality do universals have?**

Universals are general concepts. **Scholastic Realists** believed that *universals possess objective reality*, higher reality even than individual things. To medieval realists, **universals were anterior to particulars**. The more universal, the more perfect. **John the Scot** (810–877) and **Saint Anselm of Canterbury** (1033–1109) believed that universals were *res*, things. Thus the term realism. According to **William of Champeaux** (1070–1121), the individual characteristics of particular members of a species are accidental and unimportant. "Mankind" as a substance can be postulated to exist, even when there are no people anymore. (The example shows how human thinking gets out of control when it is not constantly normalized, either by observing reality, or by challenge from other minds.)

In the contemporary definition, realism is the doctrine that material things have their own independent existence regardless of an observing mind. Attributes are possessed. A realist in the common definition is someone who sticks to facts, to real, material things, in contrast to imagined ideas. For the idealist, observing mind and object are somehow connected. Truth lies in ideas rather than observables. Ideas are more real than material things. Thus, scholastic realism is a form of idealism.

The conflict about universals can be seen in the context of the general concern regarding the relationship between our ideas and the external world. How objective is our understanding of reality? In contrast to scholastic realists, **nominalists** believed that only particular things are real. *Universals are not substances, but mere names.* (Latin: nomina = names). Universals exist as ideas in the human mind. They are different from the things that they represent. They are common names for similar objects.

**Roscelin** (1050–1120) (Johannes Roscellinus of Compiegne) made the mistake of applying the concept of nominalism to the holy trinity.

Thus, he said, since only individual things as substances are real, and the reality of a whole lies in its parts, the Holy Trinity consists of three individual Gods, not of one divine substance in three personae. Ecclesiastical curse was immediate and the teaching of nominalism henceforth forbidden.

In spite of the church ban, **William of Ockham** (1290–1349) could not bring himself to accepting Scholastic Realism, because it violated the principle of *economy of thought*. Only individual things are real, because they suffice to explain reality. Universals to Ockham were mere signs, *signa*, establishing correlations between a manifold of things. It was a useless duplicity to ascribe them objective reality and being as substances.

To avoid problems with church dogma, **Ockham proposed to separate logical from theological reasoning**, accepting the dogmata of faith as concepts beyond reasoning. *He thus destroyed the union of knowledge and faith, philosophy and theology, science and religion, the common ground of facts and values.* The dual value systems that characterize our society, adherence to different values in public than private, on Sundays than during the week, have thrived ever since.

## IV. Kant and the Quantum Reality Question

*Kant:* **Our observation is of the appearance of things, not of real things; a "thing in itself" cannot be observed.**

*Quantum mechanics:* **ψ, the world not observed, is different from the phenomena observed.** *Bohr:* **We do not know the quantum reality, but only our interactions with that reality.**

The philosophy of Kant is a particularly rich source of ideas which have re-surfaced in the ontology of quantum mechanics. Some characteristic aspects of Kant's metaphysics are described in the following section as far as they provide a useful background for the parallel views that can be found.

1.

Human Knowledge extends as far as our experience, but objects of knowledge are composite, containing **a priori** and **a posteriori** elements. All knowledge begins with experience, but is not solely derived from it.

Remember Locke (the empiricist): *There is nothing in the human mind that experience has not put into it,* and the answer given by Leibniz (the rationalist): *True, there is nothing in the human mind that experience has not put into it, except the human mind itself.*

2.

We have to distinguish between the **contents** of knowledge, the matter of fact, and the **form** that it is given by human sensibility. For example, space and time are the pure forms of sensibility.

3.

Necessary and universal propositions cannot be derived from experience. They must be a priori.

"If a judgment is thought with strict universality, that is, in such manner that no exception is allowed as possible, it is not derived from experience, but it is valid a priori . . . Necessity and strict universality are sure criteria of a priori knowledge, and are inseparable one from the other." (Kant, 1929, p. 42)

Planck: *To begin with, every scientist has to take a metaphysical leap.* This is the leap onto the a priori.

4.

Space and time possess **objective reality** and **transcendental ideality**. That is, *detached from human intuition they are nothing.* They are not properties which are found in the things themselves. That is, they are created by the way we make observations. In the sense that our reality cannot be without space nor time, *observation creates reality.*

*"Transcendental ideality"* means: as far as they concern a priori modes of knowledge, or the a priori nature of reality, space and time are not real.

"Space does not represent any property of things in themselves, nor does it represent them in their relation to one another. That is to say, space does not represent any determination that attaches to the objects themselves, and which remains even when abstraction has been made of all the subjective conditions of intuition . . .

Space is nothing but the form of all appearances of outer sense. It is the subjective condition of sensibility, under which alone outer intuition is possible for us." (Kant, 1929, p. 74)

5.

All our intuition is nothing but the representation of appearance. What objects may be in themselves remains completely unknown to us. All things are appearances to us. There is no other access to them in any other way beyond their appearance. "A thing can never come before us except in appearance" (Kant, 1929, p. 286). And, in quantum mechanics: *All wave functions can never come before us, except by the observation of something related to their interferences.*

... all our intuition is nothing but the representation of appearance; that the things which we intuit are not in themselves what we intuit them as being, nor their relations so constituted in themselves as they appear to us, and that if the subject, or even only the subjective constitution of the senses in general, be removed, the whole constitution and all the relations of objects in space and time, nay space and time themselves, would vanish. As appearances, they cannot exist in themselves, but only in us. What objects may be in themselves, and, apart from all this receptivity of our sensibility, remains completely unknown to us, and not necessarily shared in by every being, though, certainly, by every human being. With this alone we have any concern. Space and time are its pure forms, and sensation in general its matter. (Kant, 1929, p. 82)

6.

At the foundation of our comprehension of reality we find various processes of unification, integrating activities aimed at the synthesis of the components of knowledge.

(a) **Sensibility** unifies the **stimuli of the senses** to **perceptions**. To do so, it uses the *pure forms of space and time.*

(b) In the next step, **understanding** unifies perceptions to **concepts**. It does so by using categories as pure forms of understanding.

(c) In the last step, **reason** guides understanding as understanding guides sensibility. It is the faculty of the highest synthesis of knowledge: it combines the concepts of understanding to conclusions. In this function it is guided by **ideas** as regulative principles.

In (b), *categories* are **pure concepts**; of unity, reality, causality, existence. They are forms that make understanding possible, but they are not part of the contents of knowledge. Since they arrange and order the objects of knowledge, they prescribe laws a priori to nature, and make nature possible.

Categories are concepts which prescribe laws a priori to appearances, and therefore to nature, the sum of all appearances. The question therefore arises, how it can be conceivable that nature should have to proceed in accordance with categories which yet are not derived from it, and do not model themselves upon its pattern; that is, how they can determine a priori the combination of the manifold of nature, while yet they are not derived from it. The solution of this seeming enigma is as follows.

That the laws of appearances in nature must agree with the understanding and its a priori form, that is, with its faculty of combining the manifold in general, is no more surprising than that the appearances themselves must agree with the form of a priori sensible intuition. For just as appearances do not exist in themselves but only relatively to the subject in which, so far as it has senses, they inhere, so the laws do not exist in the appearances but only relative to that same being, so far as it has understanding. Things in themselves would necessarily, apart from any understanding that knows them, conform to laws of their own. . . .

Consequently, all possible perceptions, and therefore everything that can come to empirical consciousness, that is, all appearances of nature must, so far as their connection is concerned, be subject to the categories . . .

We cannot think an object save through categories; we cannot know an object so thought save through intuitions corresponding to these concepts. (Kant, 1929, p. 172)

In (c), ideas are **regulative principles** of human reason. There are three classes of transcendental ideas: they are the ***concepts of the unity of the thinking subject, of the world, and of God.***

7.

*The laws of nature are laws of the understanding.* They do not apply to things in themselves, but only to appearances. They are made by the human mind. That is, human reason understands only what it produces by its own design. The order and regularity in the appearances, which we call nature, we ourselves have put in there. Thus, understanding is the faculty of rules. It is the lawgiver of nature: **The laws of physics are made by the human mind.**

8.

The only possibility for science to deal with knowledge is to rely on experience. The only possibility for science to be reasonable is to rely on a priori principles.

9.

Ideas, the regulative principles of reason are beyond the possibility of experience. No experience can be equal to any of them. They are part of thinking, not of knowing, and they are not subject to demonstrative proof or refutation. For example, it is impossible to prove or disprove the existence of God, because God is not an object of experience.

10.

Thus, the chain of synthetic reactions in forming understanding proceeds as follows:

*Data* (signals) unified by *sensibility* using the forms of space and time, give rise to *perceptions*. (Digital computers have clocks; if they malfunction, data processing ends.)

Next: *perceptions*—unified by *understanding* using the *categories* (forms) (pure concepts) of unity, reality, causality, existence—give rise to *concepts*.

Next: *concepts* are synthesized to *conclusions* by *reason*, using the *ideas*, the modes of connection. There are **three possible modes of connecting concepts**, corresponding to three classes of ideas:

the **psychological idea** (the idea of the soul),

the **cosmological idea** (the idea of the world), and

the **theological idea** (the idea of God).

Like a data-processing program, ideas are merely prescriptions.

*The idea of the soul* says: Connect all psychological appearances in such a way, as though there was a unit, the soul.

*The idea of the world* says: Connect all phenomena (observed appearances) in such a way, as though there was an underlying unit, the universe.

*The idea of God* says: You shall think in such a way, as though everything that exists has a necessary cause: God. Using the three modes, try systematically to unify all your knowledge. Thus, ideas are a program of the human mind.

As in **Eccles' theory of perception**, the mind has an integrating function. Ideas are possible to think; that is, they do not contain an inner contradiction. But they are not a matter of cognition and knowledge. Do not confuse with knowledge or cognition the possibility of thinking without contradiction. There is a temptation to do so. If you give in, reason will lead to antinomies, contradictions, and reason becomes sophistic. For example, we cannot use reason to prove or refute the existence of God.

*Critique of Pure Reason:* "I had to dissolve knowledge and understanding, in order to make room for faith."

**Thus, there are limits of knowledge.** They are found where empirical knowledge ends. Beyond these limits reason cannot achieve anything. Metaphysical ideas, like freedom, God, immortality, cannot be demonstrated, nor refuted, but one can believe in them. There are things that we can know and others we can reasonably believe in.

There can be no doubt that all our knowledge begins with experience. For how should our faculty of knowledge be awakened into action did not objects affecting our senses partly of themselves produce representations, partly arouse the activity of our understanding to compare these representations, and, by combining or separating them, work up the raw material of the sensible impressions into that knowledge of objects which is entitled experience? In the order of time, therefore, we have no knowledge antecedent to experience, and with experience all our knowledge begins.

But even though all our knowledge begins with experience, it does not follow that it all arises out of experience. For it may well be that even our empirical knowledge is made up of what we receive through impressions and of what our faculty of knowledge supplies from itself. (Kant, 1929, p. 41)

## 11.

All things must be taken in a twofold sense, namely, as appearances and things in themselves. One and the same thing can be subject to a law of nature, namely when it is an appearance, while as a thing in itself it is not subject to that law. Human beings belong to two worlds: as things in themselves they are endowed with a free will which is not subject to the laws of nature. In the world of appearances, however, their actions follow the natural laws. There is, therefore, no conflict for the morality of a human being in a nature controlled by the principle of causality. Morality presupposes freedom of the will.

Remember the dual nature of physical reality in Heisenberg's ontology: *Reality is created by two processes. One is wave-like, ruled by strict causality, and represents tendencies for becoming real. The other one is particle-like, ruled by chance, and represents transitions in observations from the "potential" to the "actual."*

## 12.

In studying nature, experience provides us with the rules and is the source of truth. In trying to determine the rules of moral actions, we cannot rely on experience. Since an element of contingency is attached to all truth based on experience alone, that which we should do cannot be derived from that which is done (Kant, 1920).

## 13.

We demand that any concept be made sensible. That is, that an object corresponding to it be presented in intuition. Otherwise the concept would be, as we say, without sense, that is without meaning (Kant, 1929).

## 14.

Theoretical knowledge is knowledge of what is. Practical knowledge is knowledge of what ought to be.

## 15.

We cannot know but think freedom. In general, there are things that we can know—that is, they have a correlate in experience—and things that we can think—that is, they are not self-contradictory. Human reason is impelled by a tendency of its nature to go out beyond the field of its empirical employment. We therefore must always be careful not to confuse the subjective condition of thinking with the cognition of an object. Our ideas, specifically, are never part of our knowledge (Kant, 1929).

16.

Truth consists of the agreement of knowledge with its object. If it does not agree with the object to which it is related, it is not knowledge (Kant, 1929). No document of truth can be found outside the realm of possible experience (Kant, 1920).

17.

*"The land of truth—an enchanting name!—is surrounded by a wide and stormy ocean, the native home of illusion, where many a fog bank and many a swiftly melting iceberg give the deceptive appearance of farther shores, deluding the adventurous seafarer ever anew with empty hopes, and engaging him in enterprises which he can never abandon and yet is unable to carry to completion."* (Kant, 1929, p. 257)

18.

*"Two things fill my mind with ever increasing admiration and reverence the more I think about them: the starry sky above and the moral law inside me."* (Kant, 1920, II, p. 174)

CORRELATIONS BETWEEN EPISTEMOLOGY
AND ETHICS

The last two quotes illustrate a phenomenon emphasized before: dealing with knowledge is not an automatic, objective, purely rational, and merely quantitative activity. Rather, it is an *emotional* issue. In addition, in Kant's philosophy, we encounter the connection between one's views of the world and of human conduct.

In Kant's definitions, a **maxim** is a rule that determines the conduct of an individual. **Imperatives** are practical laws that apply to all human beings. **Categorical Imperatives** hold generally, unconditionally.

Characteristic aspects of Kant's epistemology are (a) *the distinction between form and content of knowledge;* and (b) *the claim that all knowledge contains a priori, rational components.*

Very similarly, in **Kant's ethics**, a distinction is made between *the form of a moral law and its purpose.* Furthermore, *reason is claimed to contain a priori principles* which seek to influence human volition. As regards the **form** of moral laws, we can identify heteronomy and autonomy. Heteronomy means our will is influenced by external laws; human conduct is regulated by external principles, such as *striving for perfection, happiness, the common good.* Autonomy implies *our will is influenced by principles which originate in our reason.*

The conventional Western view has been that the human will is ruled by external principles, as human knowledge has to conform to objects. In Kant's philosophy, as objects conform to the mind, at least as far as their appearance is concerned, man's volition follows principles of his own reason.

Example of a **categorical imperative**: Always act in such a way that the maxims guiding your will can be the basis of a universal law.

## V. Positivism

The proposition *that nothing can be known about a system when observations are not made* represents a **positivistic** element in quantum physics. **If you don't see it, don't talk about it.**

Positivism in philosophy is the doctrine that human knowledge is restricted to the realm of experience, the *"positive," "facts,"* the *"proven,"* the *"practical."* It demands strict adherence to the data, avoiding anything else, in particular metaphysical principles. According to this doctrine, metaphysics and theology are preliminary stages to *"positive science."*

In the history of philosophy, one of the main proponents of positivism was **Auguste Comte** (1798–1857). His *"Law of three stages"* states that, in history, human cognition progressed from *Theology* to *Metaphysics,* and finally to *Positive Science.*

Thus, as a scientist, one should avoid questions about the *"qualities of things," "the nature of reality," "causes."* To a positivist, it is not a relevant question whether nature is causal or acausal. Neither causality nor acausality are principles of nature, because none of them can be observed. They are not scientific concepts but, rather, *poetic symbols.* The wave-particle duality is a puzzle only in the interpretations, not in the data. Thus interpretations should be avoided and everything ignored that is beyond what is recorded by the senses.

In science, one of the main representatives of positivism is the physicist **Ernst Mach** (1838–1916), well summarized by Bradley (1971): "Do not interpret; do away with all metaphysics; look at what happens; do not discuss issues like causality; do not create dogmas like that of the 'atom.' The only task of science is to predict the results of future observations from previous ones."

According to Mach (as quoted by Bradley, 1971):

> Let us look at the matter without bias. The world consists of colours, sounds, temperatures, pressures, spaces, times, and so forth, which now we shall not call sensations, nor phenomena, but simply **Elements**. The fixing of the flux of these elements, whether mediately or immediately, is the real object of physical research . . . We call *all* elements; insofar as we regard them as dependent on this special part (our body), we call them sensations. That the world is our sensation, in this sense, cannot be questioned.

A common body, such as the kitchen table, is not absolutely permanent. It is a rather stable complex of color, sound, and pressure elements. The term "kitchen chair" is no more than a compendious mental symbol for groups of sensations, and any such symbol has no existence outside of thought. Mach: *"One must not attempt to explain a sense perception. It is something so simple and so fundamental that the attempt to trace it back to something even simpler, at least at the present time, can never succeed."* It is impossible for a sense perception to be false. True and False are predicates applying to judgments and propositions, not to perceptions. Errors arise in reports by disregarding attendant and relevant circumstances. *"The combined experience of humanity down the ages is entirely insufficient to support the great scientific systems."* In this context Mach's notion arises that there is no isolated or detached element of experience. *"All that has been, can be, and could be our experience, is one."*

## VI. *Popper and Eccles: The Three-Worlds Hypothesis and Observer-Created Reality*

In 1977 **Karl Popper** and **John Eccles** put forward the thesis *that the self-conscious mind is real and exists independently of any object, notably independent of body and brain.* The starting point of the hypothesis was a specific definition of reality:

*Something is real if it can affect the behavior of a large-scale physical object.* According to this definition, all ordinary things, substances, fields, are objects of reality. They form what Popper called **World-1**. World-1 represents one level of reality.

A second level is formed by brain states. Brain states are real, because they can affect the behavior of objects in World-1, but they are different from them and form **World-2**.

In addition to the first two, *products of the human mind* form yet a different level, not fitting into either World-1 or World-2, but being real nevertheless in the sense of the stated definition. Products of the human mind include the works of poets, musicians, scientists; the contents of museums and libraries, architectural monuments, our history; and the structures we are born into, political, legal, social. *Such objects are real because, with the help of an intervening conscious mind, they have the ability to affect objects in World-1.*

Consider a poem, for example, and ask yourself the question, what exactly is it? It is more than a World-1 object—a dirty piece of paper—and is different from World-2. However, when it becomes part of the awareness of a human mind, it will be the cause of brain states—elements of World-2—that may lead to actions in World-1. Thus, poems are real.

*The striking aspect here is that products of the mind—elements of World-3 —are real only by intervention of a conscious mind. Without the awareness of a mind they are not real but have the **potential** of becoming real. Their reality is created by observation.*

Considering the special significance of mind in this process, Popper and Eccles took an additional step and postulated the *independent reality of the self-conscious mind itself.* According to this theory, mind not only is different from all the other objects of reality, but also independent of them. The theory implies that the mind may exist outside of the brain, even though it can interact with the brain and change its states, affecting World-1 objects in the process. *Thus it is believed that changes in the brain can occur that are not due to ordinary energetic causes, but to some other mechanism, such as the flow of information.*

Ordinary things are ruled by non-material quantum waves. Human beings are ruled by the non-material mind.

## VII. Skepticism

Since we do not have immediate experience of things, but only of our interactions with things, the existence of an objective reality cannot be proved. Since we can only prove our own existence and sensation, no logical law nor any experience is in contrast to the view of the skeptics that the world consists of nothing but our sensations.

One of the important skeptical movements in the history of philosophy is connected with the movement of the *Sophists,* evolving at about 500 B.C. In Russell's translation, sophists were "professors," "lawyers," "intellectuals"; the root of the term is related to the Greek word for knowledge or wisdom.

Sophists traveled from one city to another and instructed young men in practical skills that were useful to win a case in court or to succeed in debate in political meetings. To them, rhetoric was the art of persuasion based on techniques that were used to turn a given opinion into its opposite. **Protagoras from Abdera** (480–410 B.C.): *"For every object two contradictory propositions can be found."*

The skepticism of Protagoras approximately followed the line of reasoning that since all knowledge is based on perception, since each man's perceptions differ from those of the next, there is no objective truth.

Since observations are true to an observer at the moment of observation, but never generally and constantly true, perceptions are not general, but relative. In the words of Protagoras: *"Man is the measure of all things; of those that are that they are; of those that are not that they are not."* "Man" here means each individual person, not species. A person never knew of things as they were, but as they appeared to this individual in the moment of perception. Having thus established that there is no objective truth, the conclusion followed that there are no absolute moral standards, either, and that all human values are relative.

**Georgias of Leontinoi** (Sicily, 400 B.C.) carried the argument to the extreme of nihilism. In his essay "On Non-Being or Nature" he claimed that (1) Nothing exists; (2) Granted that something existed, it could not be known; and (3) Granted even that something could be known, it could not be communicated.

The skeptical movements in philosophy have shown that unproveable reality can be logically doubted. However, logical conclusions do not necessarily make sense. Logical reasoning constantly needs correcting, normalization as it were, by something other than the laws of logic. Bertrand Russell (1948): **"Skepticism is logically impeccable, but psychologically impossible."**

## VIII. Synopsis

Our views of physical reality typically affect our views of the human order that we accept. That such a correlation exists is a blessing. Without the normalizing influence of reality our thinking has a way to get out of hand, leading to concepts that make life intolerable. The discovery of the quantum phenomena has awakened mankind out of a dogmatic slumber, offering opportunities for a better world. Therefore, it is unacceptable that many of our political rulers, spiritual leaders, administrators, and teachers are scientifically illiterate.

All concepts unverified at the time of conception, and unverifiable, are *anticipated concepts*. Perhaps it is by the subtle operations enabled by the flow of probability fields that anticipated principles can enter our thoughts. Since such processes are not of the crude type of exchanges of energy, special efforts have to be made to prepare our minds to be receptive. This is a particularly important aspect of educating the young. In addition to conveying technical skills and information, the goal of education must be to engender creative minds. Every creative process makes use of anticipated principles. There is a time delay between conception and justification—turning a possible concept into a real one—sometimes requiring ages, sometimes just days.

# DEFINING A REALISTIC VIEW OF THE WORLD

In view of the complexities of the nature of knowledge it is meaningful to give a precise definition of what constitutes a **realistic view** of reality.

The search for facts is the expression of a fundamental human drive, which is the craving for knowledge. It is as basic to human nature as other needs, like the hunger for food or for love. Without the constant intake of facts, a human mind will die, like a body devoid of nourishment or affection.

Aristotle's *Metaphysics* begins: *"All men by nature desire to know. An indication of this is the delight we take in our senses; for even apart from their usefulness they are loved for themselves; and above all others the sense of sight . . ."*

The illegitimate components in the elements of knowledge are an embarrassment and a predicament. Is there anything that we can know, with certainty, clearly and distinctly? In the history of the West this problem has often been the cause of suffering, and *Divine Curse* and *Original Sin* were eagerly accepted as appropriate symbols for the state of the matter, because we have typically been driven by a desire for absolute certainty, by a craving for definitely established facts. The legendary words of **Archimedes** (285–212 B.C.), *"Give me a solid spot where I can stand, and I shall move the earth,"* are symbolic of a civilization which has always searched for solid ground in matters of a truth which can hardly be found. Thus we always wanted to **prove** the existence of God. *"I think, therefore I am,"* said Descartes, seeing *"clearly and distinctly"* many other things that later were not sustained by analysis. *"I do not frame hypotheses,"* Newton proudly asserted, at the same time claiming that the laws of nature were the same at all times and everywhere in the universe. In this way our seemingly best-established arguments and those of our convictions that at one time seem absolutely undoubtable, later on often reveal unexpected aspects, at first overlooked. In this way even the best minds of our history have frequently supported naive and uncritical beliefs in the certainty of their thinking, and few of us will escape the craving for certainty where certainty does not exist. *"Please tell me that you love me,"* we say, *"for sure, and forever."*

I define as **realistic** an attitude that is aware of the composite nature of knowledge and attempts to maintain the appropriate balance among its constituents. In thinking critically, we have to be able consciously to keep apart and to distinguish what, in our knowledge, is data (i.e., given in reality and derived from experience); what is rationality (i.e., contributed by reason); and what is the non-rational and non-empirical element involved in the derivation.

It is realistic not to confuse a single component with the whole and not to mistake for knowledge that which can merely be thought in the realm of reason or hoped for and desired in a universal realm. Historical errors in Western thinking occurred when the balance was destroyed by undue emphasis on a single component, such as experience (in **empiricism**), reason (in **rationalism**), or universal principles (in many historical movements of **idealism**).

*Without experience there is no knowledge,* but empirical data alone do not establish knowledge. *Without reason there is no knowledge,* but reasoning alone does not lead to knowledge. *Without universal principles there is no knowledge,* but universals alone, unless there is a conjunction with experience and reason, do not convey any knowledge. The cooperation of the various factors creates an effect not contained in any single one of them.

# REFERENCES

Arnheim, R. 1964. *Art and Visual Perception.* Berkeley: University of California Press.

Baggot, L. 1992. "The Meaning of Quantum Theory." *Oxford Science Publication.* New York: Oxford University Press.

Barrow, J. D., and F. J. Tipler. 1985. *The Anthropic Cosmological Principle.* New York: Oxford University Press.

Bohm, D. 1957. *Causality and Chance in Modern Physics.* New York: Van Nostrand.

Bohm, D. 1980. *Wholeness and Implicate Order.* London: Routledge and Kegan Paul.

Bohr, N. 1934. *Atomic Theory and the Description of Nature.* New York: Cambridge University Press.

Bohr, N. 1958. *Essays 1932–1957 on Atomic Physics and Human Knowledge.* New York: Wiley.

Bohr, N. 1963. *Essays 1958–1962 on Atomic Physics and Human Knowledge.* New York: Wiley.

Bradley, J. 1971. *Mach's Philosophy of Science.* London: Athlone Press.

Davies, P., and J. Gribbin. 1992. *The Matter Myth.* New York: Simon and Schuster/Touchstone.

Dobzhansky, T. 1967. *The Biology of the Ultimate Concern.* New American Library.

Driesch, H. 1908. *The Science and Philosophy of the Organism.* London: Adam and Charles Black. (The Gifford Lectures, 1907.)

Eccles, J. C. 1979. *The Human Mystery.* New York: Springer International. (The Gifford Lectures, 1977–1978.)

Eddington, A. S. 1930. *Nature of the Physical World.* New York: Cambridge University Press.

Eddington, A. S. 1939. *Philosophy of Physical Science.* New York: Cambridge University Press.

Einstein, A. 1955. *The Meaning of Relativity.* Princeton, N.J.: Princeton University Press.

Einstein, A., B. Podolsky, and N. Rosen. 1935. "Can Quantum-Mechanical Description of Reality Be Considered Complete?" *Physiology Review* 47: 777–80.

d'Espagnat, B. 1979. "The Quantum Theory and Reality." *Scientific American* (November): 2–15.

Ford, K. W. 1963. *The World of Elementary Particles.* New York: Blaisdell.

Gribbin, J. 1984. *In Search of Schrödinger's Cat.* New York: Bantam Books.

Haftmann, W. 1965. *Painting in the Twentieth Century.* New York: Praeger.

Heisenberg, W. 1952. *Philosophical Problems of Quantum Physics.* Pantheon; reprinted 1979 by Ox Bow Press, Woodbridge, Conn.

Heisenberg, W. 1962. *Physics and Philosophy.* New York: Harper Torchbooks.

Herbert, N. 1988. *Quantum Reality*. Garden City, N.Y.: Anchor Press/ Doubleday.

Herbert, N. 1993. *Elemental Mind: Human Consciousness and the New Physics*. New York: Dutton/Penguin.

Hofstätter, H. H. 1988. "Introduction to F. Souchal." *Das Hohe Mittelalter*. München: Naturalis Verlag.

Horgan, J. 1992. "Quantum Philosophy." *Scientific American* 267 (July): 94–101.

Kant, Immanuel. 1929. *Critique of Pure Reason*. N. K. Smith (trans.). London: McMillan.

Küng, H. 1978. *Existiert Gott?* München/Zürich: Piper.

Lear, J. A. 1988. *Aristotle: The Desire to Understand*. New York: Cambridge University Press.

Lorenz, K. 1966. *Evolution and Modification of Behavior*. London: Methuen.

Magee, B. 1985. *Philosophy and the Real World*. LaSalle, Ill.: Open Court.

Margenau, H. 1977. *The Nature of Physical Reality*. Woodbridge, Conn.: Ox Bow Press.

Margenau, H. 1983. *Open Vistas*. Woodbridge, Conn.: Ox Bow Press.

Margenau, H. 1984. *The Miracle of Existence*. Woodbridge, Conn.: Ox Bow Press.

Monod, J. 1971. *Chance and Necessity*. New York: Knopf.

Pagels, H. 1982. *The Cosmic Code*. New York: Simon and Schuster.

Penrose, R. 1989. *The Emperor's New Mind*. New York: Oxford University Press.

Planck, Max. 1981. *The Philosophy of Physics*. London: George Allen and Unwin Ltd., *Where Is Science Going*, 1933; repr. Woodbridge, Conn.: Ox Bow Press, 1981.

Pietschmann, H. 1980. *Das Ende des Naturwissenschaftlichen Zeitalters*. Wien/ Hamburg: Zsolnay.

Polanyi, M. 1958. *Personal Knowledge*. Chicago: University of Chicago Press; London: Routledge and Kegan Paul.

Polanyi, M. 1966. *The Tacit Dimension*. New York: Doubleday.

Polkinghorne, J. C. 1985. *The Quantum World*. Princeton, N.J.: Princeton Science Library.

Popper, K. R. 1984. *Logik der Forschung*. Tübingen: J. C. B. Mohr.

Popper, K. R., and J. C. Eccles. 1977. *The Self and Its Brain*. New York: Springer.

Powell, C. S. 1990. "Can't Get There from Here." *Scientific American* (May): 10.

Rae, A. I. M. 1986. *Quantum Physics: Illusion or Reality?* New York: Cambridge University Press.

Russell, B. 1946. *History of Western Philosophy*. London: Allen and Unwin.

Russell, B. 1948. *Human Knowledge*. New York: Simon and Schuster/ Touchstone.

Russell, B. 1978. *The Problems of Philosophy*. New York: Oxford University Press.

Scarr, S., R. A. Weinberg, and A. Levine. 1986. *Understanding Development*. New York: Harcourt Brace Jovanovich.

Schrödinger, E. 1958. *Mind and Matter.* New York: Cambridge University Press.

Sheldrake, R. 1988. *The Presence of the Past.* New York: Vintage Books, Random House.

Sherrington, C. S. 1940. *Man on His Nature.* New York: Cambridge University Press.

Sorensen, R. 1991. "Thought Experiments." *American Scientist* 79: 250–63.

Stapp, H. P. 1993. *Mind Matter and Quantum Mechanics.* New York: Springer Verlag.

Wheeler, J. A., and W. H. Zurek, eds. 1983. *Quantum Theory and Measurement.* Princeton, N.J.: Princeton University Press.

Wigner, E. 1964. "Two Kinds of Reality." *The Monist* 48: 248–55.

# INDEX